The Sensible Universe

THE SENSIBLE UNIVERSE

PHILLIP PETERSEN

Copyright © 2019 Phillip Petersen

All rights reserved

ISBN: 9781794640269

Table of Contents

PREFACE ix
INTRODUCTION xi
CHAPTER 1: HOW WE KNOW 1
    Direct Experience 2
    Logic and Reason 3
    Cause and Effect 7
    Experiment 12
    Statistics 13
    Instinct 17
    Intuition 18
    Being Told 18
    Revelation 19
    Faith 20
    Occam's Razor 20
    Teleology 21
    Imagination 21
    Models and Simulations 21
    Paradigms 23
CHAPTER 2: SCIENCE 25
    The Public Image of Science 27
    Methods of Science 28

    Physics 31

    Science, Technology and Invention 40

    Biology 40

    Evolution 41

    Genetics 43

    Environment 44

    Population 44

    Anthropology and Sociology 45

    The Mind 46

    Pseudopsychology 48

    Decisions 48

CHAPTER 3: MEDICINE 49

    Alternative Medicine 50

CHAPTER 4: ECONOMICS 52

CHAPTER 5: ANTISCIENCE, JUNK SCIENCE, PATHOLOGICAL SCIENCE AND PSEUDOSCIENCE 55

CHAPTER 6: THE PARANORMAL, THE OCCULT, THE SUPERNATURAL AND OTHER NON-SCIENTIFIC WAYS OF LOOKING AT THINGS 57

    The Paranormal 58

    Vitalism 59

    Occult and Occultism 59

    New Age 60

    Mystics 61

    The Supernatural 61

    ETs, UFOs and Other Strange Phenomena 62

    Fraud and Hoaxes 63

CHAPTER 7: RELIGION 65
CHAPTER 8: ETHICS.. 68
CHAPTER 9: PHILOSOPHY 71
CHAPTER 10: CULTURE AND THE HUMANITIES 73
    History and Pseudohistory 74
    Geography.. 74
    Language 75
    Politics 75
    The Arts... 76
CHAPTER 11: EDUCATION 77
    'Traditional' Teaching... 78
    'Progressive Education'... 78
    Postmodernism 79
    Outcome-Based Education 80
CHAPTER 12: THE MEANING OF LIFE.. 84
CHAPTER 13: CONCLUSION(S) 87

PREFACE

My thesis, of which I hope to convince you in this book, is that the universe is sensible in two senses—that it can be apprehended by the senses; and that it makes sense.

I am not claiming that only things we can apprehend with our senses exist; the tree still falls in the forest though there is no one there to see it or hear it fall (if it were otherwise, most of the universe would not exist), and there are vibrations above and below our capacity to detect as sight or sound. What I am claiming is that there is no reason to believe in any kind of transcendent reality beyond the natural, whether this postulated reality is that of idealism or is paranormal, supernatural or other occult mystery, including religion.

To put it in a nutshell, I maintain that the only real path to knowledge is verified experience. Through the efforts of thousands of deep thinkers and practical men and women over millennia, we have greatly expanded our capacity for experience and our ability to verify that experience. My plea is that we not throw away these gains by reverting to magical and other non-sensible explanations.

INTRODUCTION

'The comprehension of truth calls for higher powers than the defence of error.'—Goethe

How do we know anything? How do we know that what we believe to be the truth is in fact true? These are questions that have occupied thinking people for millennia, and ones which I will address within the bounds of my knowledge.

Where should you start if you are trying to understand the universe, life and all that? From the ancient Greeks on, a common view has been that the logical place to start is with yourself. Indeed, introspection has been the basis of a good deal of philosophy and psychology.

From the moment we are born (and perhaps even before we are born), we begin to acquire, test and utilise knowledge. One of the things a baby does early in life is to learn to distinguish self from non-self. The senses are, of course, crucial to this process. Perhaps the next stage is to distinguish living from non-living, which is an essential part of learning to manipulate the environment. Along with this goes distinguishing differences in both living and non-living things one comes into contact with. The process of developing a self-concept is then largely a piecemeal one of identifying similarities and differences between oneself and others—in one aspect, I fit into one class; in another, into another. In some ways, I am more or less unique; in others, I'm like everyone else. We will probably find exceptions to both of these and therefore modify our concept.

As we grow, we assume more and more roles which become part of our self-concept—male/female; son or daughter; brother or sister; mother or father; student; plumber, microbiologist or other job; employer or employee; etc. There are also the many other labels—smart, average or dumb; fat, average or thin; good-looking, average or ugly; tall, average or short; athletic or not. Again, there are the many personality facets others may attribute to us and which we may or may not agree with—introverted/extroverted, cheerful or sourpuss, optimistic/pessimistic, aggressive or passive, leader or follower, etc. We may identify ourselves with various groups by residence, race, religion, employment, membership of various societies and clubs.

It is important in our search for truth that we are aware of our self-image and the factors that have formed it and how this may influence our conclusions.

The pursuit of knowledge remains a major preoccupation throughout our lives. As a species, we have accumulated an enormous amount of knowledge and are continuing to accumulate it at an ever-increasing pace.

Obviously, there is also much we do not know. We can anticipate that much of this will be revealed to us in future years by the usual processes of acquisition, multiplied by building on an ever-increasing knowledge base. New technologies may at any time lead to a burst of knowledge acquisition by enabling observations, measurements or manipulations of data that are currently not possible.

Despite this, there are some things that are not merely unknown but unknowable. Given the immense scale of the universe, it is difficult to see how we could ever hope to penetrate the secrets of more than a small corner of it in our vicinity.

To all intents and purposes, the universe is to humans infinite in time and space—whether that is true or not. The trouble is that the concept of infinity is one of many that we can and must use but which essentially has no meaning for us. This may be because of the way our brains operate. Perhaps in the future our brains will evolve to such a state that they can grasp such concepts. There appear to have been equally drastic changes in our brain structures during the course of our evolution, though it is difficult to see what would drive such a process in modern *Homo sapiens*. On the other hand, perhaps the concept is basically meaningless, and we will in time come to comprehend this fact.

Perhaps it will also happen that we will be able to use computers to expand the capabilities of our puny brains— not just, as at present, by enabling computations we could never achieve by other means, but by formulating and testing concepts and models beyond our brainpower. I must admit to being sceptical. Experience to date with artificial intelligence does not give much cause for optimism that the super-intelligent machines beloved of science fiction are possible, let alone imminent.

Again, there are things which are logically undecidable. As Gödel showed 86 years ago, arithmetic is one of these. That is, there are statements within arithmetic that cannot be proved true or false within the system. As Turing further showed, computers cannot decide whether a mathematical statement is true or false.

Further, many problems, though theoretically decidable, are computationally intractable; the resources required for their computation are so far off the scale as to be, for practical purposes, (? forever) unreachable. The travelling salesman problem is one of these, where a salesman must make calls in a number of cities and the problem is to minimise the total distance travelled, while ensuring that he returns to his starting point.

Then there are what Weinberg has called 'trans-scientific' questions.[1] An example he gives is whether the amount of electromagnetic energy to which we are normally exposed increases the risk of cancer. He points out that a study of the spontaneous mutation rate in mice exposed to 150 millirem of X-rays would require eight billion mice in order to show an increase of one-half percent. The experiment, while theoretically possible, is obviously impractical in real life.

There is, thus, the known, the unknown and the unknowable. Being able to decide between these cases may well be said to be the beginning of wisdom.

---

[1] Weinberg, AM. Science and transcience. *Minerva* 10:207-222, 1972.

CHAPTER 1

HOW WE KNOW

'To give a reason for anything is to breed a doubt of it.'—William Hazlitt

Dictionary definitions of 'fact' include: '1. Something known with certainty. 2. Something asserted as certain. 3 Something that has been objectively verified. 4. Something having real, demonstrable existence.'

There have been those, such as Giambattista Vico, who believed that the only things we could really know with certainty were things, like mathematics, that we had created. It is certainly true that only a Divine Creator could know all possible facts.

Most of what we regard as fact comes more into the third definition. In our everyday experience, we are continually testing and verifying what we believe to be facts, and frequently changing our minds if our first impression proves to be incorrect. Much of our knowledge is, in reality, a best guess—or our acceptance of someone else's best guess. The more it has been tested, the more secure in it we can be, but absolute certainty is a mirage.

This is difficult for many people to accept. They demand the certitude of 'eternal truths' as espoused by dogmas of various kinds. Their 'facts' are those of sense 2. Asserting something as certain does not make it so.

Science is an extension of our common-sense approach to knowledge. 'Facts' are accepted as such after attempts to disprove them have failed. This makes all scientific knowledge tentative—a failure in the eyes of the many who, despite their everyday experiences, desperately seek for certainty.

In truth, science is no different to the common-sense attitude of reasonable people of ever being on the alert for anomalies that may conflict with their presently held views of reality. Unfortunately, many have accepted the misinterpretation of the old saying that the exception proves the rule. 'Prove' here originally meant 'test', not 'establish the truth of'. This is the way in which science and all sensible people operate.

What we must guard against is self-deception and wishful thinking. As Thomas Gilovich has pointed out in *How We Know What Isn't So*, three tendencies we must strongly guard against are to:
- misperceive random data and see patterns where there are none;
- misinterpret incomplete or unrepresentative data and give extra attention to confirmatory data while drawing conclusions without attending to or seeking out disconfirmatory data;
- make biased evaluations of ambiguous or inconsistent data, tending to be uncritical of supportive data and very critical of unsupportive data.

Direct Experience

'What we observe is not nature itself, but nature exposed to our mode of questioning.'—- Werner Heisenberg

From the time we are born (and perhaps before) much of our knowledge of the world is based upon our own direct experience of it. As Miguel Cervantes said, experience is also 'the universal Mother of Sciences.'

Unfortunately, our senses can deceive us. We have all experienced mirages and other optical illusions. There are also auditory illusions. A tonal stimulus of high pitch when presented straight ahead appears to originate above a tonal stimulus lower in pitch.

Other animals also quite literally see the world differently to the way we do. We see only three primary colours and derive all the rest by mixing these, while octopi only see blue, birds see four or five colours, bees see ultraviolet, blue and yellow, and snakes have some vague colour perception but also see infrared. Some have senses we do not: sharks and rays can detect electric fields and use this ability to hunt fish; migratory birds have internal compasses; bees can sense the polarisation angles of ultraviolet rays from the sun; many species have specialised organs to detect subtle traces of behaviour-stimulating pheromones.

Clearly, we cannot always trust our senses and there is more to reality than what our senses can detect. This does not mean, as some have argued, that we can never believe our senses or that there are mystic forces we will never be able to understand.

Logic and Reason

'Like every other human endeavour, logic is just a patchwork quilt whose patches do not meet very well, and which are continually being torn up and restitched.'—I Hacking

Logic and reason are the processes by which we make sense of what our senses tell us. Sense data lead to an awareness of an occurrence but inference is necessary for perception. Useful inferences are thought processes that integrate past and current experience. Because of limitations of our sense data and our knowledge base, we need to constantly check our conclusions. In effect, we put up a working hypothesis and attempt to verify it by obtaining other pieces of information—by direct observation or by experiment. At the same time, we employ falsification criteria to check that our assumptions are correct. Thus, inference leads to certainty—perhaps not of the kind of 'eternal truth' but the more useful guide to successful behaviour.

The two main forms of inference in Greek logic are deduction and induction.
Deduction takes the form:
All humans are mortal;
All Greeks are humans.
Therefore, all Greeks are mortal.

This system of logic is based on the idea of universals and amounts to placing things in sets and examining the relationships between the sets. These sets are 'crisp' sets; a given thing must either belong or not belong to the set. Deductive logic can really only deal with one kind of uncertainty—nonspecificity, lack of informativeness. It can tell us to which crisp set something belongs. (Birds and bats both fly but birds have feathers and lay eggs while bats do not have feathers and bear live young.)

Fuzzy set theory is an attempt to deal with vagueness by assigning values between 0 (definitely does not belong) and 1 (definitely does belong). We can use it to determine the extent to which something is true. Fuzzy logic has produced washing machines that do everything but load the laundry and push the button and promises remarkable technological advances from software that predicts the stock market based on the daily news to cars that drive themselves.

Deductive logic is said to produce a priori knowledge, i.e., it does not depend for its authority on the evidence of experience. In fact, if the premises used (such as *All humans are mortal*) are not based upon experience, the deduction will produce nonsense.

The inductive process is one of generalisation of empirical observations. Inductive inferences invite doubt because these observations may not be entirely suited to generalisation.

There has been an ongoing dispute for centuries about which logical system science uses to arrive at its conclusions.

One view is 'hypothetico-deductivism'. As described by physicist Richard Feynman in his lecture 'Seeking New Laws', 'In general we look for a new law by the following process. First, we guess it. Then we compute the consequences of the guess to see what would be implied if this law that we guessed at is right. Then we compare the result of the computation to nature…to see if it works.' [2]

In Popper's version, hypotheses cannot be confirmed, only found to be tested and not yet falsified.

Against this, Whewell maintains that 'the records of mankind offer no single instances of any great physical truth anticipated by mere guesses and conjectures.'[3]

Pierce agreed and maintained that a hypothesis must be invented by a 'retroductive' inference of the form:

The surprising fact, C, is observed;

But if A were true, C would be a matter of course,

Hence, there is reason to suspect that A is true.[4]

---

[2] Feynman, Richard. 1965. *The Character of Physical Law*. Cambridge, Mass.: MIT Press

[3] Whewell, William. 1831. 'Review of J Jerschel's *Preliminary Discourse on the Study of Natural Philosophy (1830).' Quarterly Review* 90:347-407

[4] Peirce, Charles S. 1960. *Collected Papers of Charles Sanders Peirce*. Vols.7-8 edited by Arthur W Burks (1960). Cambridge, Mass.: Harvard University Press

Unlike the hypothico-deductivist, the inductivist denies that consequential testing is sufficient for confirmation; he requires inference from the data to the hypothesis as well. Unlike the retroductivist, the inductivist does not claim that this inference is a unique explanatory one.

According to Peirce, induction is 'the process by which we collect a *General Proposition* from a number of *Particular Cases*...The Inductive step consists in the *suggestion* of a conception not before apparent.' This conception unites the facts and renders them capable of being expressed by a single law.

He also said, 'When our Conceptions are clear and distinct, when our Facts are certain and sufficiently numerous, and when the Conceptions, being suited to the Facts, are applied to them so as to produce an exact and universal accordance, we attain knowledge of a precise and comprehensive kind, which we may term *Science.*'

Whewell claims that the inductive discovery of an empirical law typically involves an inference, or series of inferences, to a new property shared by members of a class, followed by the generalisation of this property to all members of the class, including its unobserved members[5]. This certainly seems to have been the way in which such great scientists as Kepler and Newton operated.

There is also another scheme of inference which appears to correspond more closely with what we do in everyday life. Though elements of this appear in Greek thought, it is most developed in the Nyāya system, developed in India around 400 AD. As given by Rogers and Jain[6], this can be set out as follows.

There is fire on the hill;
(statement of the working hypothesis)
there is smoke there;
(citation of evidence)
wherever there is smoke, there is fire, as in the kitchen;
(invoking a general principle with a specific example)
tatha
(given all the specifics above, it follows that)

---

[5] Whewell, William. 1858. *Novum Organon Renovatum*. London. John W Parker.

[6] Rogers JM and Jain MK. 1993. Inference and successful behaviour. *Quarterly Review of Biology*, 68: 387-407

there is indeed fire on the hill.
(conclusion, statement accepted).

According to this view, knowledge is a formalism of past experience and derives its authority from nothing else. Understanding is a matter of being able to interact successfully with the event or object at hand. The most efficient way of acquiring knowledge is through trial and error. In the case of illusion, there is superimposition of memory on perception and can be dealt with by the same iterative validation or invalidation that we use in other cases.

Nyāya deals with uncertainty simply by pointing out the unreasonableness of doubting as long as successful behaviour results.

It is likely that a combination of the methods of Nyāya and analytical reasoning is the key to successful behaviour in both philosophy and science.

What logic can certainly do is to alert us to certain logical fallacies in our own thinking and in the arguments of others. A common one of these is the argument from ignorance. Basically, this occurs when one argues that a thing must be true because it cannot be proved false. This is, of course, a patent absurdity. You can't prove to me that there are not little green men on Mars who eat cheese every day and twice on special days, but that does not make the claim true.

Related to this is the divine fallacy. Anything we can't understand or provide an explanation for must be due to God (or aliens, or paranormal or supernatural forces).

Another common fallacy is begging the question, when one assumes what one claims to be proving. In classical form, the argument for intelligent design of the universe could be put:

We see the perfect order of creation in the universe.
This order demonstrates the presence of a supernatural intelligence in its design.
Therefore, God exists.

The premises of this argument presuppose the existence of a supernatural intelligent designer, i.e., God. For the argument to succeed, it would first be necessary to demonstrate that there is perfect order and that this could only arise from intelligent design.

These three fallacies illustrate misuse of deduction. A common misuse of induction is confirmation bias, the tendency to notice and look for what confirms one's beliefs, and to ignore, not look for, or undervalue the relevance of what contradicts them. This is something we must all be wary of both in ourselves and in others.

One should perhaps also make a brief mention of paradoxes. A well-known example, 'I always lie', is a modern version of an old Greek paradox. The paradox is that, if the statement is true, it must also be false. The solution is simply to realise that the statement is false. This is true of most paradoxes.

Cause and Effect

>*'"For example" is not proof.'*—Yiddish Proverb

The concept of causality is one that is central to our experience. Very early in our lives, we discover that certain actions cause certain other events to follow. Conversely, we come to expect that a particular event has been caused by a preceding event.

However, the mere fact of one thing happening after another does not prove that the first event was the cause of the second event. This mistaken idea, the so-called post hoc fallacy, is the basis for many superstitions and erroneous beliefs. Many sportsmen have lucky charms or rituals based on a better performance on some occasion; these are undoubtedly as effective as primitive peoples beating on drums to bring back the sun during an eclipse.

Controlled studies are often an effective way of dealing with this problem. These involve the comparison of a control group to an experimental group. The groups must be identical except for the factor that is being measured. The study should be double-blind; i.e., neither subjects nor experimenters know who belongs to which group. The clinical trial is a good example of this. Unfortunately, while appropriate controlled evidence is demanded before new pharmaceutical agents can be marketed, the same demand is not applied to many new devices and surgical procedures, or to many so-called natural products. While it is not feasible to test every assumed association in this way in everyday life, we can, and often do, set up less rigorous trials similar in principle.

Another fallacy that makes evaluation of cause and effect difficult is the regressive fallacy—the failure to take into account natural and inevitable fluctuations. This accounts for the dozens of ineffective arthritis remedies on the market and the testimonials that many of these can produce for their effectiveness.

The apparent success of many of these products is due to the fact that their use is commenced at a time when the patient is feeling particularly bad and the improvement experienced is due to the natural fluctuation in severity that is a feature of the disease. It is a common occurrence that many sufferers go from one remedy to another in this manner as the current one proves unable to deal with the inevitable exacerbation.

As an illustration of the difficulties in proving causation, I will discuss in some detail the problem of assigning causation in disease, and especially in infectious disease.

Koch's postulates are frequently used to establish causation of disease. Koch's perspective of disease, the so-called germ theory of disease, dates back some six millennia to the Egyptian perspective of disease as occurring when a parasite finds an environment suitable to life within the body of a host and being due to the activity of the parasite. The Egyptian perspective was refined by Fracatorius, with his view that communicable disease is caused by a living contagium vivum which is transmitted by direct contact, by fomites, and through the air.

The Koch perspective opposes not only the Indo-Assyrian-Babylonian perspective of disease as being caused by malign outside influences which are either spiritual or inanimate, and the 'balance' perspectives such as the Empedoclean (health is a state in which external and internal environments are in harmony), the Hippocratic and Galenic (disease is due to disorders in the vital fluids of the body), and the Paracelsian (disease occurs when changes in the external or internal environment produce undesirable changes in the host beyond those which the host is capable of compensating for), but also the 'New Medicine' assumption that the causes of most illness are to be found in environment, life-style and emotional/sensory balance.

While not incompatible with Metchnikoff's perspective that infectious disease occurs when a parasite's ability to invade exceeds the capacity of a host to defend itself from attack, nor yet with Ehrlich's viewpoint of disease as a state in which the internal environment becomes inimical to the host due to the presence of various toxins, or even with Nuttall's perspective of infectious disease occurring when effects due to an invasion by a parasite are not sufficiently neutralised by processes evoked in the host's body fluids, or with Deny's concept that resistance to infection depends on cooperation between humoral and cellular defence mechanisms, it does not take these into account.

Although it has long been recognised that infection is much commoner than disease, the Koch perspective ignores this fact. A more modern perspective sees infection as occurring when an organism capable of establishing itself at a site in or on the body of a host is not prevented from doing so by the activities of the host or of other organisms present at the site, and infectious disease as occurring when the activities of an infecting organism become such as to damage the host. In contrast, the Koch perspective maintains its naive invariant relationship between various bacteria and specific diseases.

The Koch postulates are intended to prove that a specific microbial species is the specific cause of a specific disease entity. Unfortunately, things are not always that clear-cut; not only can the same organism produce a somewhat different clinical picture in different cases, but different organisms can produce indistinguishable disease states.

This tends to make nonsense of the first of Koch's postulates, which may be stated in generalised modern form: the microorganism must be present in every case of the disease. What frequently happens is that the disease state is diagnosed on the basis of the isolation of the pathogen. In fact, it is the corollary of Koch's postulates—that is, when an organism has been shown to cause a certain disease, any patient from whom that organism is recovered is therefore suffering from that disease—which has formed the basis of clinical microbiology for nearly a century. It is somewhat ironic that this is so, for in fact Koch utilised his postulates in only two of his investigations—anthrax and tuberculosis. In his 'Investigations into the Aetiology of Traumatic Diseases', his methods were purely histological; and in his other great work on cholera, he was forced to conclude that it was not possible to provide a proof by using his postulates.

The second of Koch's postulates was that the microorganism must be grown in pure culture. This insistence on pure culture technique was invaluable, but the standard method of studying microbes in isolation is unlikely to produce dramatic results of the kind achieved by Koch in his classical studies of anthrax and tuberculosis. There are numerous references to differences between organisms grown in culture and in vivo.

Koch's third and fourth postulates were that the specific disease must be reproduced when a pure culture of the microorganism is inoculated into a healthy host; and the microorganisms must be observed in, and recovered from, the experimentally diseased host. These two steps have frequently proved stumbling blocks in attempts to apply Koch's postulates, as Koch himself experienced in the case of cholera.

In 1980, I pointed out that the processes of infection and infectious disease are sequential, with each step representing a metastable state. The change from one state to the other may occur either as a catastrophic change or as a steady transition, this being so because any change in any relevant variable depends not only on that variable and its interaction with other variables but also on the actual state of the system and its previous history. The result of each step (elimination of the parasite, commensalism, latent infection, subclinical infection, chronic infection, acute infection, kind and degree of injury, exit of parasite) is determined by an interplay of parasite, host and environmental factors.

I suggested a new set of postulates of pathogenicity more suited to the current situation as follows. The organism must: (1) either be shown to be producing infection at the biological site in question or produce infection in a specific cell system replicating the conditions prevailing at the relevant site; and (2) either be shown to be producing effects which constitute, or can be quantitatively correlated with, the symptoms of the condition, or be shown to be capable, under the conditions prevailing at the site, of producing such effects; (3) evidence of a quantitative relationship between such effects and the activity of the organism must be obtained; (4) it must be demonstrated further that the organism is inhibited in its capacity for producing these effects by agents mitigating the symptoms of the condition; (5) presumed cause and effect throughout the sequence of events leading to the disease state must be shown to be temporally related, for example, by the fact that a regression of current causal variable A on past, present and future values of effect variable B should show significant coefficients for future, and perhaps present, values of B, but insignificant coefficients on past values of B, while regressing B on past, present and future values of A should yield significant coefficients on past, and perhaps present, values of A and insignificant coefficients on future values of A.

Similar criteria for evidence of causality rather than association between observed phenomena and a disease were given by A Bradford Hill[7]. These were: (1)

---

[7] Hill, A.B. The environment and disease: Association or causation? *Proc.R.Soc.Med* 58:295-300, 1965

strength of the observed association; (2) consistency of the observed association; (3) specificity of the observed association; (4) the temporal relationship of the association; (5) the presence of a dose-response relationship; (6) the biological plausibility of the association; (7) coherence of association (the cause-effect interpretation should not conflict with known pathology of the disease in question); (8) experimentation—does preventive action based on some supposed association in fact prevent the outcome in question?

It is the last criterion which allows us to find necessary causes of disease—though they may not be sufficient causes. As we shall see, experiment is an essential step in obtaining and verifying knowledge in many other areas.

One illustration of this is the demonstration by Addison that the disease which now bears his name was due to a disease of the adrenal cortex. His finding that all fatal cases of the disease showed a diseased adrenal cortex on autopsy was supplemented by the demonstration that removal of the gland from experimental animals led to the disease. The clincher was that an active principle, which could be isolated from functioning glands but not from diseased glands, was found to relieve symptoms when injected into diseased persons.

Many diseases are currently thought to be multi-factorial. This may be so but, as with tuberculosis, there may, in many cases, be one overwhelming factor which is necessary, though not sufficient, to cause the disease.

The cause(s) of various diseases are often established epidemiologically, using statistical methods. Unless and until such criteria as Hill's are applied, these connections must remain speculative. As an illustration, John Snow was able to show, back in 1854, that an outbreak of cholera in London was associated with water from a particular well. He concluded that the disease was caused by a specific agent with the property of multiplying in human beings, passing from the sick to the healthy and probably having the structure of a cell. At the same time, Dr Farr, the superintendent of statistics, showed statistically that the death rate from cholera varied inversely with altitude above sea level. Using the prevailing 'miasma' paradigm of contagious disease, he interpreted the data as demonstrating that cholera was caused by high atmospheric pressure. Even Snow's abortion of the cholera epidemic by barring access to the contaminated water did not destroy official acceptance of this theory though it would clearly fail most, if not all, of Hill's criteria.

One of the most valuable lessons science has learned, and one that may well be applied to everyday life, is that it is frequently more relevant to consider under what circumstances A causes B, rather than to dogmatically conclude that A causes B under all circumstances.

A spurious assumption of homogeneity is a cause of error, not only in establishing causation but also in proving the efficacy of intervention. A good example of this is the history of anticoagulant therapy. A large controlled trial found benefit for the use of anticoagulant therapy in myocardial infarction but was later found to have had a defective assignment mechanism. Properly randomised trials showed no advantage. However, other research showed that it was beneficial for patients with arrhythmias, congestive heart failure or other complications but useless or even harmful for other patients.

A caution perhaps needs to be issued that cause and cure do not always go hand in hand. There are instances where an effective treatment for a disease exists even though it is not known what causes the disease, and there are instances where the cause is well known but no effective treatment exists.

Experiment

'To solve a problem it is necessary to think. It is necessary to think even to decide which facts to collect.'—Robert Hutchins

'The process of experimental science does not consist in explaining the unknown by the known, as in certain mathematical proofs. It aims on the contrary to give an account of what is observed by what is imagined.'—Francois Jacob

The experiment is at the heart of the scientific method and legends of critical experiments abound. However, the answer provided by a decisive experiment may not always be the final one. For instance, the controversy between Fuller Allbright (who had evidence that parathyroid hormone acted on the kidney) and JB Collip (who found it acted directly on bone) was supposedly settled by the fact that the hormone showed an effect in animals whose kidneys had been removed. Eventually, it was found that both men were right, and the hormone acted at both sites. The interpretation of the results of an experiment can be even more important than the way in which it is designed and performed.

In science as in everyday life, observations are frequently dismissed or neglected not only because of their low credibility or indirectness or triviality, but also because they contradict current beliefs—which may or may not have been adequately verified. As G Myrdal put it, 'generally speaking, we can observe that scientists in any particular institutional setting move as a flock, reserving their controversies and particular originalities for matters that do not call into question the fundamental system of biases they share.'[8]

Reductionism, the belief that an effective way to understand a complex system is to understand its component parts, is a cornerstone of science. This approach can be extremely productive as long as one remembers that a complex system cannot be represented simply by the addition of all its component parts. Biology cannot be reduced to physics and chemistry, though physics and chemistry can certainly help us understand biology.

Experiments do not always turn out the way they were planned. The good researcher will not merely dismiss these as failures but will attempt to find the reason for the unexpected result. Much has been made of the role of luck in scientific discovery, one of the most famous being Alexander Fleming's accidental observation of the inhibition of staphylococci by a mould, which led eventually to the discovery of penicillin. The truth is, of course, that luck is only useful to those who use it; or, as Pasteur put it, 'chance favours only the prepared mind.'

Statistics

'Statistics are like a bikini. What they reveal is suggestive. What they conceal is vital.'—A Koestler

Statistics deals with probabilities, and probability deals with dissonance—pure conflict. It can tell us the odds of an event happening and may influence our decision to accept or reject a hypothesis.

It can protect us against the clustering illusion—the intuition that random events occurring in clusters are not really random events. Statistically, there is a 50% chance of getting four heads in a row when we toss a coin and a better than even chance that any given neighbourhood will have a statistically significant cluster of cancer cases.

---

[8] G Myrdal. Objectivity in social research. London: Duckworth, 1969

Epidemiologists call the clustering illusion the Texas sharpshooter fallacy, after the story of the Texas sharpshooter who shoots holes in the side of a barn and then draws a bullseye around the bullet holes. Politicians, lawyers, environmental activists and some scientists have noted individual cases of disease and then drawn boundaries suggesting a cluster. A statistician might explore a possible connection between a cluster of leukemia cases and use of pesticides by conducting a case-control study, in which the exposure of children with leukemia (or of their pregnant mothers) is compared with that of children of the same age, sex and other characteristics but without leukemia. He or she could refine the process by determining which pesticides are typically elevated where leukemia cases tend to cluster. Of the thousands of cancer clusters investigated, not one has convincingly been identified as due to an underlying environmental cause.

Unfortunately, statistics are grossly abused in the hands of many epidemiologists. It would seem that any combination of 'exposure' and disease, regardless of biological implausibility, or even without any underlying hypothesis, is fair game for calculating relative risks, odds ratios or proportional hazards.

Perhaps the commonest abuse surrounds the concept of risk factors. Journals and newspapers are full of reports such as one that claimed to show that 'a small risk of coronary heart disease due to baldness may exist.'[9] What should have been said is that the data suggested a small independent association between baldness and subsequent heart disease. If this were confirmed, the appropriate response would be, not for bald men to rush out and get hair transplants, but for scientists to investigate a possible linkage and biological mechanism (e.g., testosterone).

Many of these reports are contradicted by later reports that find no association. It is no wonder that the public has become cynical and tends to disregard even valid findings or to select only those that suit them.

A prime example of spurious associations is smoking. It is well established that smoking is bad for your health, but some of the effects attributed to it are patently absurd. Particularly suspect are some of the claims made for ill effects of passive smoking, which was at one time even claimed to be more deadly than direct smoking (the justification for this peculiar finding was that the smoker was getting

---

[9] Herrera, CR and Lynch, C. Is baldness a risk factor for coronary heart disease? A review of the literature. *J. Clin. Epidemiol.* 43:1255-1260, 1990.

directly inhaled smoke, while those in the vicinity were getting 'slipstream' smoke; how the smoker manages to avoid inhaling 'slipstream' smoke was not explained). So extreme were some of these claims that the World Health Organisation felt compelled to issue a disclaimer. Nonetheless, sizeable judgments are still made for people making outrageous claims for damage caused by second-hand smoke—including even one for pneumonia supposedly caused by passive smoking.

The situation is further compounded by the attempt to settle the confusion created by conflicting reports by the use of meta-analysis. To expect that a simple pooling of studies with not quite the same methodologies and not quite the same populations and of very uneven quality will make sense of conflicting studies is irrational nonsense. Also, many investigators indulge in 'data dredging'—searching all conceivable subgroups for an effect, often with computer assistance. The result is almost invariably the demonstration of a completely spurious effect.

This is not to say that meta-analysis, like other statistical methods, cannot be extremely useful. When properly done, it can indeed sometimes make sense of conflicting reports and can reveal patterns that would not otherwise be detected; for instance, meta-analysis showed that streptokinase lowers the death rate for heart attack patients treated within twelve hours of their first symptoms but increased the mortality of those treated 72 hours after those symptoms.

The way statistical epidemiological data are presented can also often be confusing. Thus, it is often stated that a woman has about a 1 in 8 chance of getting breast cancer. This is, in fact, the cumulative lifetime risk, assuming that she lives to at least 95, and obscures the fact that her risk of getting breast cancer in any given year is less than 1 in 100,000 in her twenties, rises to a peak of 1 in 193 at age 77 and then falls to about 1 in 1,000 when she passes 95.

So-called 'vital statistics', on which national and international comparisons of various diseases are made and trend analyses performed, are so fundamentally flawed that any conclusions must be considered highly suspect. These statistics are taken from causes of death recorded on death certificates. Efforts to standardise coding systems (which regularly change) have not been matched by efforts to standardise diagnostic criteria, with errors of unknown but large magnitude. This problem is compounded by the increasing use of encoding directly into computer storage, with no 'paper trail' capable of being checked.

Another compounding factor is the inaccuracy of censuses. Even in the United States, undercounts of 10-20 percent in some regions are known to occur. This will result in falsely elevated morbidity and mortality rates for these areas.

The message of all this is that statistics is an essential tool but one that is horribly misused, even by scientists. A statistician once analysed all the articles involving statistics (most of those appearing) in several issues of one of the most respected journals in infectious diseases and found the statistics in almost all of them flawed.

No wonder the general public is confused and can be persuaded that using lotto numbers that have not been chosen in recent drawings or that have come up more frequently than expected is going to win them a fortune. This 'gambler's fallacy', the mistaken idea that the odds of something with a fixed probability increase or decrease depending on recent occurrences, is only one common misapprehension of statistics.

People also regularly confuse probability and possibility. The possibility of my having four cups of coffee for breakfast is 1 (i.e., certain) but the probability is near enough to zero.

Again, have you ever thought of the incredible odds of you (or me, or any specific individual) being born? Firstly, there are the remote odds of a specific sperm fertilising a specific ovum. Then there are the chances that events in the womb did, or did not, produce epigenetic changes. Go further back than that and think of the odds that your parents even met and had reproductive sex. And so on back through the generations. If you could do the calculations, you'd soon enter the area of improbability. And yet you were born and are alive to read this!

A possibility theory has been developed that cannot resolve the type of uncertainty that probability theory can but works with nonspecificity (basic ambiguity) and measures confusion by assessing how easily an event can happen, its potential. Thus, while probability theory can tell us how probable it is that a certain value of a liver function test indicates hepatitis, possibility theory can tell us how possible it is that it is really due to stomach disease.

Dempster-Shafer theory combines features of both possibility theory and probability theory by allowing an 'uncertain' category as well as a straight 'yes' or 'no'.

Instinct

There are certainly things we do instinctively—i.e., without any activating thought or the need for experience. But such things are probably far fewer than we usually imagine. For instance, we imagine that the baby instinctively suckles the breast; in actuality, the baby instinctively suckles and only learns by experience that the breast is the place where it should suckle if it wants a feed.

There was a time when it was held that humans did everything by reason and non-human animals did everything by instinct. Those who held this belief can never have observed either humans or other animals. The human baby certainly sucks without thinking about it or needing to be told or shown. And much animal behaviour is predicated on experience, including imitation and being actively taught by parents.

One of the commonly observed instances of this is the learning of predatory behaviour and techniques by felines. Mothers bring prey to the kittens, release them and encourage the kittens to catch them, preventing the prey's escape and demonstrating correct techniques. As with humans, there are good and bad learners and good and bad teachers, and I have even known mother cats not averse to delivering a good belt to the head to a kitten not paying attention.

Despite this, people are suckers for the myth that animals (and, some believe, humans when they allow their instincts to come to the fore) inherently know all sorts of wise and wonderful things.

A famous instance of this is the widespread knowledge that migratory birds can find their way unerringly around the world through some kind of magic instinct. The truth is that a combination of very keen eyesight, grid recognition skills and the presence of magnetite in their ears enables quite precise navigation but it's basically a game of follow the leader.

In the late 70s and through the 80s and into the 90s, there were several reports of animals apparently being instinctively guided to plants which were either curative for the illnesses or contained substances missing in their diet. More careful studies suggest that most, if not all, of these instances involve learning by trial and error and social observation, rather than instinct or 'body wisdom'.

Intuition

Intuition is a supposed faculty that enables us to know things without conscious, logical thought processes. In this sense, it undoubtedly exists, but exactly what is it and how does it work?

The common explanation is that it is a mechanism for directly tapping our unconscious, or subconscious, mind. The whole concept of the subconscious is a vague proposition which merely acknowledges that much of our mental activity is carried out below the level of conscious awareness. This is not to say that there is anything radically different in these thought processes; rather, they consist of the associative, analogical processes that are our normal method of thinking, rather than the sequential logic we insist characterises 'conscious' thought.

Intuition in this sense can be very valuable but it can lay no special claim to absolute truth. Intuitive conclusions can be brilliant insights; they can also be garbage. An example of this is the cop's supposed 'blue sense' that enables him to detect impending danger and solve a case by inspired hunches. This intuition, insofar as it exists, is undoubtedly due to honed skills of observation and free associative thinking and is by no means infallible.

Still, sudden insights do occur—in composers, inventors, painters, poets and scientists. In all cases I have come across, they occur in a state of relaxation, when the person is not consciously seeking a solution. Frequently, motion is involved, as in walking or travelling in some mode of transport, but the various techniques of mediation may also bring the same results.

Being Told

'An expert seldom gives an objective view. He gives his own view.'—Morarji Desai

If our knowledge was restricted to what we ourselves experienced and verified, we would know very little. We rely heavily on what other people, experts and non-experts, tell us. How do we know that what they tell us is correct? The first requirement is that it make sense.

Almost every day, we are presented with testimonial evidence. This may be something as simple as your friend telling you, 'I used Brand X and it worked'; or it may be a glowing account in an advertisement of how such and such a program produced effortless weight loss; or someone earnestly telling the world on TV how they were abducted by aliens.

How do you know which of these to believe? I would maintain that the correct attitude is, at best, suspended judgment for any claim that cannot be, or has not been, validly tested. In the case of your friend's recommendation, you might wish to buy the product and try it yourself. With the weight loss program, I'd want to see the results of controlled trials. And for the alleged alien abduction, I'd want, as a first step, an interrogation by a skilled cross-examiner.

What if the person making a claim is an authority in the field? One must, of course, take such claims seriously but I would still require valid evidence. It may be that the evidence is beyond my capacity to judge because of my lack of background knowledge in the field. I would still want to know that the claim is testable, and has in fact been tested, before I gave it credence. And I would want to assure myself that any apparent lack of clarity was due to my lack of knowledge and not to muddled thinking on the author's part.

The point of all this is that no belief becomes true on the basis of who believes it and, to the limits of one's abilities, one should test all things.

One needs to be especially alert to the process of communal reinforcement, whereby a claim becomes a strong belief through repeated assertion by members of a community. As has been repeatedly shown in history, the fact that everyone in a community believes something does not make it true.

Revelation

A claim for absolute truth is made in many religions on the basis of a supposed divine revelation. A more modern version of the phenomenon is the claim by some to have had knowledge revealed to them by aliens or the spirits of past notables.

Every such instance I have examined—including all the revealed religions—contain such implausibilities and inconsistencies that one can only conclude that either the persons to whom they were revealed were deluded or their informants were not the omniscient, or at least very knowledgeable, entities they were supposed to be.

Revelation is intrinsically an unsatisfactory (invalid) means of knowing because it is incapable of validation or reproduction.

Faith

Faith is a non-rational belief in some proposition. Religions often try to make a virtue of faith. To me, such an attitude is even more irrational. Any belief that is founded purely on faith—i.e., cannot be supported by argument and is, in fact, contrary to the sum of the evidence—is, simply, not a sensible belief.

Some religionists claim that such scientific theories as evolution amount to statements of faith. This is pure nonsense. Scientists accept or reject such views based on the evidence for or against them. They may speculate beyond the evidence, but such speculation, if it is to be scientifically useful, must be capable of validation. The same applies to common sense experience.

Occam's Razor

'Plurality should not be posited without necessity.'—William of Ockham

'Seek simplicity and distrust it.'—Alfred North Whitehead

'While Occam's razor is a useful tool in the physical sciences, it can be a very dangerous implement in biology. It is thus very rash to use simplicity and elegance as a guide in biological research.'—Francis Crick

Occam's razor has been much used in philosophy and science and even in everyday life. Basically, at least as it is used today, it implies that the fewer assumptions (especially of causative agents) that have to be made the more likely a view is to be true and the more useful such a view will be.

Thus, an atheist may argue that having God as the creator of the universe introduces an unnecessary assumption since, if the universe needs a creator, why does God not need one, and if God does not need one, why does the universe? It would appear more reasonable (and more in accord with Occam's razor) to accept that the universe is self-existent.

The quote by Whitehead above is a good statement of how Occam's razor should be used. A good example of this from medicine would be tuberculosis. Before Koch proved tuberculosis was due to a bacterium, there were a myriad of causes attributed. When medicine accepted *Mycobacterium tuberculosis* as the cause of tuberculosis, a tremendous advance in the understanding of the disease occurred, resulting in effective measures for prevention and treatment. However, the matter did not quite stop there. It was now possible to look at subsidiary factors with a new eye and to determine those that contribute to an individual's likelihood of contracting the disease.

Crick's work with DNA was another outstanding example. Without his simple, elegant explication of the genetic code, we would be no further advanced than Mendel. And yet we now know that things are not quite as simple as his picture implied; we now know a single gene may do different things (or nothing), depending on its relationships with other genes and with the environment.

Teleology

'Teleology is a lady without whom no biologist can live; yet he is ashamed to show himself in public with her.'—Ernest Wilhelm von Brücke

Teleology, at least as used here, means the presence of design in nature. For example, honeyeaters have long slender beaks designed to reach the nectar in the bottom of flowers.

Such a view seems a common sense one but, if accepted, implies a designer—whether God or a nebulous Mother Nature. Evolution offers an alternative explanation, which will be expanded on later.

Imagination

'Imagination is more important than knowledge.'—A Einstein

Einstein is perhaps overstating the case, but it is certainly true that it is impossible to acquire such knowledge as that of relativity without a good dose of imagination. The trick is to subject our imaginings to reality checking.

Models and Simulations

'Whereas a good simulation should include as much detail as possible, a good model should include as little as possible.'—J Maynard Smith
'The purpose of models is not to fit the data but to sharpen the question.'—S Karlin

'Ecology is awash with all manner of untested (and often untestable) models, most claiming to be heuristic, many simple elaborations of earlier untested models. Entire journals are devoted to such work and are as remote from biological reality as are faith-healers.'—D Simberloff

Models and simulations are used to make sense of a fluid world where categories are not mutually exclusive. Formal analysis is not suited to this task because it breaks the world into mutually exclusive categories and lumps distinct phenomena under a single heading because of formal similarities.

Models differ in their generality, realism, precision, manageability and understandability. There is a trade-off among generality, realism and precision[10], but where a model stands with respect to these qualities depends on the context in which it is used, how it is used, the stage of an investigation, and the state of the science.

We can attempt to increase realism by adding independent variables, by adding new variables that mutually affect each other, by adding a new link between variables already present (but thereby reducing generality), relaxing simplifying assumptions (but thereby decreasing precision), or by restricting the domain of application (again reducing generality).

Most of the models we work with are partly true and partly false. 'We may strengthen our confidence in the implications of some assumptions by using ensembles of models that share a common core of these assumptions but also differ as widely as possible in assumptions about other aspects'.[11]

However, before any model can be accepted as representing even an approximation to truth, it must be tested qualitatively and quantitatively on the basis of predictions that do not depend strongly on any particular values of input parameters. In other words, models must be judged on the basis of their assumptions and of their degree of fit to the data.

Many investigators scan a phenomenon with simplified models and verify their results with simulations.

---

[10] Levins, R, and RC Lewontin. 1985. *The Dialectical Biologist.* Harvard University Press, Cambridge.

[11] Levins, R. 1993. A response to Orzack and Sober: formal analysis and the fluidity of science. *Q. Rev.Biol.,* 68: 547-555

This has been the approach taken by many traffic engineers. The result has been a lot of knowledge gained about how traffic moves or doesn't move, but with few practical results—largely because problems are as much political, economic and sociological as engineering.

Paradigms

> *'An important scientific investigation rarely makes its way by gradually winning over and converting its opponents: it rarely happens that Saul becomes Paul. What does happen is that its opponents gradually die out and that the growing generation is familiarised with the idea from the beginning.'*—M Planck
>
> *'Facts do not organise themselves into concepts and theories by being looked at; indeed, except within the framework of concepts and theories, there are no scientific facts but only chaos. There is an inescapable a priori element in al scientific work. Questions must be asked before answers can be given. The questions are all expressions of our interest in the world, they are at bottom valuations.'*—G Myrdal

As an illustration of the importance of perspectives, consider the discoveries, consequent on the invention of the microscope, that bits of cork are made up of tiny compartments, that pieces of meat are made of tiny elongated units, that leaves of trees are made of small rectangular units—all interesting, but not very useful, facts until the suggestion was made that perhaps all living things are made of these tiny units.

Again, it is unlikely that astronomers before Copernicus sat around in their coffee breaks discussing the remarkable fact that the sun moves around the earth, but this perspective of the universe so blinded them that they were apparently completely unable to see the numerous celestial objects discovered in a very few years after the Copernican view became accepted.

The 'Copernican revolution' is a good example of what Thomas Kuhn called a paradigm shift. Such paradigm shifts usually occur when anomalies are found that cannot be explained by the existing theories or do not fit the current models.

Such new concepts may be resisted, but Max Planck was a little too pessimistic. Opponents don't have to die, only retire from their positions as heads of university departments, editorships of journals, etc. And they can sometimes be circumvented by self-publication and other means. On the whole, though, inertia is a powerful force and it can take centuries for new ideas to be accepted and utilised.

Paradigm shifts are not always bloody. A gentle, but highly significant, paradigm shift occurred in neurology in the 1860s, when the community of neurologists switched from classifying neurological diseases on the basis of the physiological principle of the distinction between sensory and motor function in the nervous system to the eclectic approach giving primacy to pathological or clinical descriptions, that is still used today.

A valuable new paradigm often arises by breaking down the dualistic approach that characterises much of our thought—nature versus nurture, wave versus particle, for example. It will also often go beyond the linear thinking that constitutes much of our formal attempts at deriving knowledge and perceive new associations. Finally, of necessity, it will successfully integrate knowledge obtained by reductionist investigations of myriad components.

Perhaps the most important paradigm is the personal one; people who see the world as basically comprehensible, manageable and meaningful are usually the happiest and healthiest.

CHAPTER 2

SCIENCE

'We do not suggest that science invented intellectual honesty, but we do suggest that intellectual honesty invented science.'—J Erikson
'The object of all science, whether natural science or psychology, is to coordinate our experiences into a logical system.'—A Einstein
'The whole of science is no more than a refinement of everyday thinking.'—A Einstein
'Science is a balanced interaction of mind and nature.'—S J Gould
"One thing I have learned in a long life: that all our science, measured against reality, is primitive and childlike—and yet is the most precious thing we have.'—A Einstein
'In the continuing war against death, disease and poverty, the major weapon of our species is science. To not see that is to have closed eyes.'—Julian Tobias

What is Science for?

As stated in the quote from Tobias above, science can have great practical benefits and is, in fact, the best means we have of achieving these benefits.

But much of science is not pursued for any immediate practical benefit—though such fundamental science may result in discoveries that eventually produce unimagined benefits. The story of Faraday's response to being asked at the Royal Institution what use his discovery of magnetic induction was ('What use is a newborn baby?') is often used to illustrate this. (Personally, I like his response to the same question put by the Chancellor of the Exchequer even more: 'I don't know, but one day, sir, you may be able to tax it.')

Perhaps the greatest use of science, however, is, as adumbrated in the first two of the quotes from Einstein above, as a refinement of everyday thinking that allows us to coordinate our experiences into a logical system.

Of course, in science as elsewhere, what we think is truth may merely be majority opinion, but scientific 'truth', because of the way it is obtained and tested, is more likely to be actual than 'truths' obtained by other means.

Except in its demand for absolute intellectual honesty, science is valueless. This is both a strength and a weakness. Despite the opinions of some modern (or postmodern) sociologists, properly performed science is devoid of cultural and personal biases. Scientists must, however, have values if we are not to see a repeat of the enlistment of science in carrying out the Nazi atrocities.

Science is not a harmless intellectual pastime. It enables us to control as well as to observe. We must reject the 'technological imperative' and decide as a world community what use we will make of scientific and technological discoveries. Perhaps the 'softer' sciences like anthropology, psychology and sociology will help provide answers to the problems that the discoveries of the 'harder' sciences like physics, chemistry and biology raise.

Unfortunately, many sociologists, in concert with historians, philosophers, feminists and others, have revealed a real and deep hatred of science, arguing that science is a construct of culture, biased to its deepest roots (especially against women) and subjective. This is clearly nonsense: science is objective, consistent, consilient and predictive—none of which can be said for the opposing disciplines.

The supposed weakness of science in being subject to change is, in fact, its strength. Science is not bound by culture; it is continually testing the current orthodoxy to see if it can be improved—something we all do at our most rational. But any proposed alternative must also be tested to the highest standards of evidence and inference.

One aspect in which science is less than objective is in its reporting of results; researchers are loath to submit negative outcome studies for publication, even though such studies may sometimes be more valuable than ones with positive outcomes. This situation is greatly exacerbated by the tendency of the media to only publish scientific (especially medical) stories with positive outcomes—and to frequently proclaim the most trivial advance as a breakthrough (though it must be admitted that some researchers are not guiltless in giving this impression).

A difficulty with the pursuit of modern science is that it is increasingly expensive and heavily dependent on government grants. The grant application process is not ideal and may result in very worthy projects losing out to others less worthy. Also, because of ever-increasing demands in other areas, funds supplied tend to be decreasing in real terms. Allied to this is an increasing demand for accountability, with economic benefits being the prime criterion applied. To ensure public funding for science, research institutes must engage themselves with social and economic objectives. The people in power delegate their trust to specialists, who may not always merit or respect that trust.

Yet, whatever waste there is in the process, science more than repays the investment. In developed countries, scientists and engineers working in research and development make up between 30 and 80 of every 10,000 people employed but are responsible for some 50 percent of the growth in the economy.

For many people, the trouble with science is that, as Steven Weinberg once put it, 'The more the universe becomes comprehensible, the more it seems pointless.' As I hope to show, this ain't necessarily so.

Despite its failings, to my mind, science is the greatest invention of the past two millennia, and it is an invention that is still being improved upon.

The Public Image of Science

'It is not necessary for our politicians and civil servants to be professional scientists, but it is necessary for them to understand what science is.'—George Porter

Many people, especially humanistic intellectuals and New Agers, look down on science as the prime example of linear thinking. Nothing could be further from the truth; as shown above, imagination and associative thinking are integral components of science.

Some people are turned off by reports of fraud in science. However, such instances are quite rare and their effects minimal.

Others are repelled by the mathematics in which so much of science is expressed. This is largely a deficiency of education, which has resulted in the loss of the tradition that incorporated mathematics into the background of most serious thinkers.

Despite these attitudes, overall the view of science held by the public is very positive. A poll conducted by the Roper Center for Public Opinion Research at the

University of Connecticut in 1996 found that 83% of respondents agreed that 'science elicits satisfaction or hope and excitement or wonder'; 78% that 'science will solve most or some of society's problems'; and 75% that 'funds for pure research are a good investment'.

At the same time, movie scientists, from evil monsters to the merely insane, from bumbling nerds to stalwart heroes, still largely inform public perceptions of the real thing.

Methods of Science

> *'A good theory makes not only predictions, but surprising predictions that then turn out to be true. (If its predictions appear obvious to experimentalists, why would they need a theory?)'*—
> Francis Crick

Theory construction is the basic method of science. That is, a principle or set of principles is derived which can be used to explain, organise, unify or make sense of some range of phenomena. To be scientific, a theory must be empirical, falsifiable and possess predictive power.

In the quote above, Crick is expressing a common view of students of science—the value of novel predictions. These occur when a theory makes predictions of phenomena not previously known but which are subsequently found to be true; or when it explains previously unexplained phenomena.

One reason such novel predictions are highly regarded—other than that they can lead to a whole new field of inquiry—is that they are felt to guard against the introduction of ad hoc hypotheses. In point of fact, science does frequently make use of ad hoc hypotheses, but these are run through a sequence of tests, as a result of which some will be transmuted into legitimate scientific hypotheses.

Unfortunately, students of science seldom submit their theories to the same sort of empirical testing used in science.

However, as Franklin has remarked: '*Among the most important kinds of experiment are those that refute a well-confirmed theory or those that confirm an implausible theory. It is an experimenter's hope to find such unexpected results.*'[12]

A basic feature of the scientific method is the need for absolute objectivity. Unfortunately, conflicts of interest are common. Resnik[13] defines conflict of interest in science as follows:

'A scientist has a conflict of interest if a) he is in a relationship with another scientist or member of the public requiring him to exercise judgment in that other's service and b) he has an interest tending to interfere with 1) the proper exercise of judgment in that relationship or 2) his ability to fulfil his obligations to that person in his role as a scientist.'

His suggested steps for dealing with this problem include avoiding such financial incentives as stock, insuring the integrity and reliability of the peer review system, and applying peer review to expert witnessing, as well as the usual requirements for avoidance and disclosure.

A blatant instance of corporate pressure in science was revealed when the *St Paul Pioneer Press* reported that more than a dozen scientists received US$156,000 from the tobacco industry to write letters to scientific journals disputing the carcinogenic effects of second-hand smoke.

Another is a survey of scientists working at research intensive universities in the US, which showed that more than half of such scientists who received gifts of research material from pharmaceutical companies or biotechnology companies reported that the donors expected to exert influence over their work, including review of academic papers before publication and retention of patent rights for commercial discoveries.

Similarly, in 1998, the plant and microbial biology department of the University of California, Berkeley negotiated an exclusive research relationship with Novartis Agricultural Research Institute that allowed the company to review graduate student and postdoc work before publication—without consulting the students.

---

[12] Franklin, Allan. 1990. *Experiment, Right or Wrong*. Cambridge: Cambridge University Press.

[13] Resnik, D B. 1998. Conflicts of interest in science. *Perspectives on Science* 6(4): 381-408

The commercialisation of university research leads to a diminution in the free flow of ideas, a focus on more applied projects and serious conflicts of interest. The ideal situation is vastly increased government support for curiosity driven basic research and a mechanism to commercialise any discoveries made in this way.

Conflicts between open science and commerce currently abound. A good example is the stifling of research through the patenting of gene sequences. Another is the field of blood substitutes, where much of the research was initiated and driven by the medical and life sciences research communities, with many thousands of papers already published, before commercial, often venture capital supported, interests realised the size of the potential market. Since then, it has happened that extensive animal experimentation has been reported on a product or formulation only identified by an undecipherable code name, and untoward reactions observed during human clinical trials were for a time known only through hearsay or through the financial press.

Intimidation of researchers by special interest groups is also a not uncommon occurrence. For example, a researcher who published results of a study of multiple chemical sensitivity was attacked by special interests, and a large corporate entity confronted with a billion-dollar damage suit because of the public health hazard of a marketed chemical product indulged in peer review corruption.

The genuine inquirer's concern for truth is sometimes described as the scientific attitude. It is not true that all scientists have this attitude or that only scientists have it. However, it is the attitude that makes science possible. Unfortunately, this critical analytical thought, vital to an accurate view of the world, faces a continuing struggle to maintain itself against a deeply embedded predisposition to believe in magic.

Non-scientific theory builders are often wilfully blind to prosaic evidence that inconveniently contradicts them. They worship inappropriate abstraction at the expense of common sense and practical wisdom. Their arguments are often flavoured with a good dose of utopianism.

Science is largely concerned with interpreting observed facts, and scientists do sometimes misinterpret their observations. Sometimes, too, they are human enough not to perform the experiments that will refute these misinterpretations or to cling to their misinterpretation when the experiments have been performed.

Much of science involves mathematics, both in the testing of its hypotheses and in the formulation of its results. But, in what sense is mathematics a science?

Modern mathematics is distinguished largely by abstract, unifying ideas and formal structures and involves little of the empirical testing for which science is noted.

Physics

'Experiment alone can decide on truth but the axiomatic basis of physics cannot be extracted from experiment.'—A Einstein

The very word, physics, implies something solid—the physical world. But much of what physics deals with is not something we can touch, though we can observe its effects. For all that, physics is the cornerstone of our appreciation of the material world and of our interactions with it. And yet even physicists have surrounded much of the achievements of modern physics in advancing our knowledge of the universe with an almost mystical aura that clouds our understanding.

Physics is a highly mathematical science. Some of the formulae of physics are derived from the approach, supposed by some logicians as the standard method, of performing experiments and deriving formulae from the observations. Many more, however, have been obtained as a result of theory and then confirmed experimentally.

Either way, since Galileo and Newton, it has been the empirical, mathematical approach which has characterised physics. Many metaphysicians following Newton tried to establish his laws of motion (or their own versions of these) non-empirically, but only historians and philosophers remember their names. Unfortunately, some modern theoretical physicists are not too different from these metaphysicians in their approach.

The fundamental formulae of physics have been tested, proved and utilised extensively but, in many cases, there is still doubt and confusion about their actual 'real' meaning. Many of them embody operational definitions. Take, for example, the mathematical statement of Newton's second law of motion: $F = ma$. Here, force is the mysterious quality that makes an object depart from its state of rest or motion, and mass is the quality a body has of resisting this force.

In looking at the results of physics, it is often necessary to differentiate between the mathematics and the interpretation of the mathematics. The mathematics may have been extensively tested, but this does not always provide incontrovertible evidence that the interpretations contained in the model, and therefore the model itself, are correct.

Hacking concluded that astrophysics does not qualify as a proper natural science but must rather be seen as model making.[14] It is true that models play a large role, but experiments enter prominently. Examples are the refutation of the Lyttleton-Bondi electrical universe theory, and experimental corroborations of aspects of the theory of relativity.

Astronomy can generate models that make successful predictions, that are easy to understand and that are easy to learn. Because of the then impossibility of obtaining measurements, Greek, Arabic and early European astronomy was strictly a mathematical subject and such physical arguments as were employed were regarded as merely persuasive. However, it is increasingly becoming possible to make actual measurements and astronomy and astrophysics are more and more coming to resemble other parts of physics. Despite this, our view of the 'macro' universe still depends on interpretations of defined entities that may not be correct.

Physics has often been regarded as the model of incontrovertible rigour in the empirical sciences but there is much about the foundations of the subject that is still incompletely understood.

When it was first expounded nearly a century ago, the theory of relativity was regarded as incomprehensible to all but an esoteric inner circle, and many non-scientists still view it this way today. Yet, it has been tested time and time again and, so far, there have been no discrepancies. For instance, the effect of gravity on time not only has been shown to exist but the global positioning system must take account of it. And yet, what does it all mean?

The theory of relativity has many parts—some more incomprehensible to non-physicists than others.

Perhaps the thing most people think of first when they think of relativity is the famous equation: $e = mc^2$, where e = energy, m = mass and c = the speed of light.

As everyone knows, this provided the basis for the development of nuclear energy and nuclear weapons. It expresses the fact that a small amount of mass can be converted to a huge amount of energy. It also indicates that energy can be converted to mass—though it would require a tremendous amount of energy to produce an appreciable amount of mass. Looking at it another way, energy can

---

[14] Hacking, Ian. 1983. *Representing and Intervening: Introductory Topics in the Philosophy of Natural Science.* Cambridge: Cambridge University Press

provide (an extremely small) resistance to a force. This equation also required a combination of the laws of conservation of mass and energy.

Another interesting thing about this equation, which seems to have been lost in the general excitement, is the involvement of the speed of light as the constant in the relationship. Why? I would suggest that it has to do with the fact that light is electromagnetic radiation.

Electromagnetic radiation consists of an electric field and a magnetic field perpendicular to each other and to the direction of propagation, and includes, in order of decreasing frequency (or increasing wavelength), gamma rays, X-rays, ultraviolet radiation, visible light, infrared radiation, microwaves and radio waves. In Maxwell's formulation, they are described in terms of differential calculus of three spatial dimensions and time.

Einstein also found it necessary to include time (and use differential calculus) in his analysis of two inertial frames moving with respect to one another. In 'Euclidean space', the length, l, of a line can be specified by the invariant: $l^2 = (\Delta x)^2 + (\Delta y)^2 + (\Delta z)^2$, where $\Delta x, \Delta y, \Delta z$ are the changes in the x, y and z directions respectively.

However, when two inertial frames move with respect to one another, this does not hold and the invariant becomes: $s^2 = (\Delta x)^2 + (\Delta y)^2 + (\Delta z)^2 - c(\Delta t)^2$, where s is called the interval, c = the speed of light and Δt = the length of the time interval. Note that the term involving time is rather different from the spatial dimensions. And there's the speed of light again!

The speed of light has another important place in the theory. Einstein's postulate that the speed of light is a constant to any observer has been amply confirmed. What is so special about the speed of light?

In his general theory of relativity, Einstein went further and considered the case in which masses were involved. He expressed this in the form of an equation, which can be given, in its simplest form, as $s^2 = Ac(\Delta t)^2 + B(\Delta x)^2 + C(\Delta y)^2 + D(\Delta z)^2 + Ec(\Delta t)(\Delta x) + Fc(\Delta t)(\Delta y) + Gc(\Delta t)(\Delta z) + H(\Delta x)(\Delta y) + J(\Delta x)(\Delta z) + K(\Delta y)(\Delta z)$, where the coefficients, A...K (together known as the metric tensor), contain the masses of the bodies that have to be taken into account.

Einstein's theory explains gravity not as a mysterious force acting at a distance (as in Newton's theory) but as the effect of mass on space-time. Now, this model neatly explains all the so far observed phenomena, but is it anything more than a model and what does it really mean?

To attempt to decipher this, let's look at quantum mechanics. Planck postulated that the total energy of a model collection of oscillators in a blackbody giving off radiation could only have certain discrete values. His calculations fitted the facts exactly.

Einstein deduced from this that individual oscillators, such as vibrating atoms, are each characterised by a certain discontinuous set of energies, or quanta, and went on to show that the energy of a quantum is given by the equation: $\varepsilon = h\nu$, where ε is the energy of the quantum, h is Planck's constant and ν is the natural frequency of the oscillator.

He went on further to show that light behaves, not only as a wave but as if made up of particles he called photons. His mathematics and those of others produced a theory which has shown an ability to predict unexpected phenomena that can be confirmed by observation and experiment with uncanny accuracy.

The heart of quantum mechanics lies in the simple two slit experiment, in which a beam of photons is shone on a barrier in which two slits have been cut, and the light passing through them is recorded on a photographic plate. When only one slit is open, there is one line on the plate. When both slits are open, the pattern shows interference between the two beams—the light acts like waves. When photons are beamed at the apparatus slowly, one by one, the pattern that builds up shows wave-like interference—as if the individual particles, beamed at different times, were able to influence each other.

The standard interpretation of this, often called the Copenhagen interpretation because this was where Bohr, its chief protagonist, worked and argued out his vision with his collaborators, is the principle of complementarity—that physical entities can display properties that are fundamentally incompatible. In the case of light, so this principle states, one can observe particle properties or wave properties but not both simultaneously. This is more or less a simple statement of experimental fact but many physicists, including Bohr and Heisenberg, went further and insisted that light could not in fact exist as particle and wave simultaneously—an assumption that goes far beyond the experimental facts.

In fact, David Bohm, through a refinement of de Broglie's pilot wave model, showed (some twenty years later) that, mathematically at least, they could. Basically, what he did was to acknowledge that particles move. By simply including the positions of the particles of a quantum system as part of the state description of that system and allowing these positions to evolve in the most natural way, the entire quantum formalism, including the uncertainty principle and quantum randomness, emerges.

The uncertainty principle states that we cannot simultaneously measure the position and the velocity of electrons (which can also exhibit both particle and wave properties). At first sight, this would appear to be nothing more than a logical limitation of measurement; it is, after all, logically impossible to precisely measure both the location and velocity of any moving object, no matter how large. But Heisenberg was implying more than this; he said that a particle could not simultaneously *have* a defined momentum and position; and, further, that subatomic particles could never truly be said to be in one place but were merely more likely to be in one place than another. Bohr went even further than this and maintained that particles did not have properties before they were observed.

Thus, Neils Bohr and Werner Heisenberg gave us a perplexing theory that not only left apparent contradictions unresolved but declared, in advance, that any attempt to resolve them is foredoomed to failure.

Einstein would have none of it and, together with Boris Podolsky and Nathan Rosen, came up with the so-called EPR thought experiment which theoretically showed a means by which it might be possible to accurately determine both position and momentum.

Nearly thirty years later John Bell proved, by simple arguments, that if the two identical particles at the heart of the EPR experiment cannot affect each other then it is impossible to obtain the results observed in the actual experiments.

Bell was a strong supporter of Bohmian mechanics and used it in interpreting the double slit interference experiment:

'Is it not clear from the smallness of the scintillation on the screen that we have to do with a particle? And is it not clear, from the diffraction and interference patterns, that the motion of the particle is directed by a wave? De Broglie showed in detail how the motion of a particle, passing through just one of two holes in a screen, could be influenced by waves propagating through both holes. And so influenced that the particle does not go where the waves cancel out but is attracted to where they cooperate. This idea seems to me so natural and simple, to resolve the wave-particle dilemma in such a clear and ordinary way, that it is a great mystery to me that it was so generally ignored.'[15]

---

[15] J S Bell, *Speakable and Unspeakable in Quantum Mechanics*, p 191

In 1982, Alain Aspect carried out an experiment which confirmed Bell's arguments—two identical particles moving within the same system do affect each other.

So far so good, but here is where physicists become mystics and maintain that the experiments prove the existence of some sort of mysterious connection between particles that can operate instantaneously over astronomical distances, and that the act of measuring one particle forces certain properties (which did not previously exist) on a second particle.

These interpretations are correct but not in the mystical way they are usually expressed. Physicists have bewitched themselves with the mystery of it all, neglecting the obvious that has been staring them in the face—light is an electromagnetic radiation, with an associated field. The fields of any two 'particles' of electromagnetic radiation (particles of matter, such as electrons, behave analogously) interact with each other and with those of macroscopic objects. In the experiments, performing a measurement of any kind on one of the particles affects its field and therefore that of the other particle. Both particles had properties before the measurement, but these properties are changed by the measurement. (This is one of the basic limitations of science—one can never perform a measurement without, to some degree, altering the thing being measured).

The standard explanation is made even stranger by the fact that it exists beside quantum field theory. Heisenberg and Pauli propounded this theory back in 1929. It treats all particles as being a bunching up of a field.

In the case of light and other electromagnetic radiation, this makes sense of Einstein's derivation (which has been verified experimentally) that the mass of a particle increases with velocity so that, at the speed of light, it becomes infinite. This means that photons, since they travel at the speed of light, cannot have mass. Which again means that they are quantised wave forms with associated fields.

But, if this is true, how can gravity affect light? Isn't gravity an attractive force producing an acceleration in a mass? Though this is the way physics still mainly uses gravity, there are currently three competing views that tend to be used as the need arises. These are: Einstein's view of gravity as a distortion of the space-time continuum; quantum gravitation, in which gravity is caused by an exchange of particles (gravitons); and the existence of gravity waves. Actually, all these competing views can be subsumed within a picture based on a melding of Bohmian mechanics and quantum field theory. So too can the other fundamental forces—

the electromagnetic, the weak nuclear and the strong nuclear (without the need for the myriad of postulated particles).

It should now also be abundantly clear that space-time is not some mysterious property of the universe but merely a construct expressing the fact that, in the words of Heraclitus, 'only change is constant'. Applying David Hestenes' geometric algebra to a three-dimensional space existing in time allows all the necessary derivations of both relativity and quantum mechanics, without the need for introducing such oddities as negative time.

Why should the speed of light be inviolable? If we look again at the equation, $e = mc^2$, we can see that this implies that the ratio of energy to mass is also a constant. It is the constancy of this ratio, plus the fact that energy can only be gained or lost in discrete amounts, that is at the heart of both relativity and quantum theory and unites the two.

Another apparently well verified part of relativity is the so-called dilation of time—i.e., to a stationary observer, a clock on a moving object appears to run slow. This has been taken to mean that time actually varies with motion, but does it mean any more than any clock (i.e., method of measuring time) we can conceive of will be affected by motion. It must be remembered that, since the speed of light is constant, at least for light a dilation of time implies a dilation of distance.

Similar remarks could be made for the effect of gravity on time.

Time is a somewhat slippery concept, due largely to the fact that we use it in so many different (but related) senses (I have found 18 different dictionary definitions and there may well be more). For the purposes of this discussion, the following will suffice: a nonspatial continuum in which events occur in apparently irreversible succession from the past through the present to the future.

Much is sometimes made of the fact that many of the equations of physics are apparently reversible. However, in reality, what happens always depends on the initial conditions and it is this, plus the fact that there is no such thing as a truly closed (or isolated) system (except the universe itself), that makes time truly irreversible.

The physical law that embodies this irreversibility is the second law of thermodynamics: entropy increases. But what is entropy?

Many definitions of entropy have been given. Probably the most commonly known is: a measure of the randomness, disorder or chaos in a system. Others are: a measure of the capacity of a system to undergo spontaneous change; and the tendency of the energy of a closed system, including that of the universe itself, to become less able to do work with the passage of time. In fact, entropy can be defined as a function of any set of macroscopic variables by the logarithm of the volume of the subset of phase space of the microscopic variables on which these macroscopic variables take a given value.

As I have said, one of the problems of physics is that the mathematics, which can usually be tested directly, are often confused with the model, which is not necessarily proven by proving the mathematics. A related difficulty is the tendency to draw all manner of unwarranted philosophical implications from physics. Entropy, a convenient mathematical expression of the relationship between microscopic and macroscopic variables, is an example of this.

Another good example is the philosophical position that chaos theory negates determinism. This position is based on a confusion between determinism and predictability; the fact that some physical phenomena may be unpredictable does not mean that they are not deterministic. This also applies to the probability descriptions of such particles as electrons in atoms; the fact that we cannot, at any given moment, precisely identify the position or trajectory of a particle probably reflects our ignorance more than a real lack of these attributes.

Many modern interpretations of probabilities by physicists seem to amount to a fundamental misunderstanding of probability. If something has a probability of 0.1, it has a one in ten chance of occurring, yet many theories (including Feynman's formulation of quantum theory in terms of a sum over histories) seem to interpret this as meaning it one tenth occurs.

Many physicists also seem to have a basic problem with language; they talk of multiple universes. The universe is the composite of everything that exists; you can only have one everything.

The currently accepted view of the creation of the universe is that it began in a Big Bang. There is considerable evidence for this point of view, but what does it really mean? Whatever existed at the moment of the big bang was, at that time, the universe. Some physicists maintain that before the Big Bang there was nothing and that space and time were created along with matter, energy and all the other bits that make up the universe. Despite our familiarity with the concept of zero, the idea of nothingness is so much against our experience that it is impossible to conceive. There are certainly no known physical laws that allow something to be created out of nothing, and plenty that disallow this. Note that introducing a divine creator does not get around this problem, because pre-creation this God *was* the universe.

Within the bounds of our current knowledge, Stephen Hawking's idea that the universe is completely self-contained, having no boundary or edge, seems the most palatable. This is a natural consequence of an expanding universe. Note that an infinite universe must also, of necessity, be expanding. In effect, then, the universe is infinite in space and time (or space-time) and is characterised by the continual creation of space and time.

Interestingly enough, Hawking is also a firm believer in the Big Bang. How does he reconcile these two contradictory positions? As far as I am aware, he has made no attempt to. One possible answer (perhaps the only one) is that Big Bangs are events that happen within a pre-existing universe.

Perhaps the extreme example of runaway speculation in physics is superstring theory. This is too bizarre to bother describing here. Suffice it to say that it is untested and basically untestable and actually, though it purports to provide a quantum explanation of gravity, explains nothing.

Another nonsensical concept is that of multiple or parallel universes. By definition, the universe is everything there is. For this reason, too, God could not have created the universe as something outside himself (or herself). At all times, God must have been the universe and the universe must have been God.

Also, if God was (and therefore is) the universe, the Devil, if he exists, must be at least part of God. If that doesn't make much sense to you, I can only say that the Christian church's account makes much less sense. According to this, God made the Devil one of the highest of the angels, but he rebelled against God and was therefore thrown out of heaven. Would an all-powerful God make such an imperfect creature? Why would anyone who had it so good rebel—especially when it would have been obvious that he couldn't win? Why then would God send the Devil down to earth to see how many men he could convince to join him?

In the same vein, it could reasonably be asked why, when he could apparently make perfect beings, did God create creatures that could (indeed, according to some parts of the scriptures, *must*) sin and then commit them to hell if they sinned? That makes as much sense as if I were to make, say, a table, then declare it was evil because it had wobbly legs and throw it in the fire. Since sin is disobedience to God, if there is no God, there is no sin. This does not mean there is no right or wrong. Ethical and moral principles can be derived without resorting to 'divine commands'.

One of the attractions of religion is that it may provide some meaning to life through the promise of paradise (with or without the threat of hell). Of course, this assumes that life has any meaning—a proposition that is nowhere near as inevitable as some believe. On the large scale, however, life may fulfil some purpose in the universe (intended or not), since life appears to be the sole means whereby a kind of reversal of entropy can be achieved; without it the universe could, perhaps, be even more chaotic and fall rapidly into complete disorder.

Science, Technology and Invention

'The science of today is the technology of tomorrow.'—Edward Teller

'A patent, or invention, is any assemblage of technologies or ideas that you can put together that nobody put together that way before. That's how the patent office defines it. That's an invention.'—Dean Karmen

Much of what is called science could more properly be called technology—the practical application of science. An invention is most likely to be useful if the inventor is familiar not only with the technology but with the science behind it and if he or she uses scientific methods to test his ideas.

On the other hand, the English geneticist, Sir Paul Nurse, has revealed that the philosophical works of Sir Karl Popper have informed the practice of his scientific research activities.

Technology has invaded biology, to the extent that discoveries in some aspects of biology may be due to an evolving technology that owes more to engineers than to biologists.

Biology

Physics and biology may seem miles apart but not only are many concepts and facts of physics essential to the study of biology, but Rudge has shown that the philosophy of experiment developed in the context of high energy physics can be extended to include examples from evolutionary biology[16].

Evolution

'Natural Selection is not Evolution. Yet, ever since the two words have been in common use, the theory of Natural Selection has been employed as a convenient abbreviation for the theory of Evolution by means of Natural Selection, put forward by Darwin and Wallace. This has had the unfortunate consequence that the theory of Natural Selection itself has scarcely ever, if ever, received separate consideration.'—Ronald Fisher

Evolution is accepted as a fundamental postulate by scientists but is largely ignored, misunderstood or rejected by the public at large.

There are those who maintain that humans and their society are continually evolving and that there is always progress—even if of the two steps forward, one step back variety. It is certainly true that humans and their societies continually evolve, but those who equate this with progress totally misunderstand evolution. Evolution is concerned with adaptation to the current and changing environment, not with progress towards some distant goal. Evolution does not provide a non-religious meaning for life. Teilhard de Chardin was wrong.

Fisher (quoted above), along with Haldane and Wright, combined Mendelian genetics and Darwinian evolution into the modern evolutionary synthesis and helped found population genetics. Genetics and evolution are now likely to be discussed together.

It is important to realise that not all genetic differences between species are due to natural selection. An often-quoted example is the difference between the alpha chain of haemoglobin in humans and gorillas, which consists only of one having glutamic acid where the other has aspartic acid at site 23. It is difficult to believe this arose by natural selection and that the two species have different varieties of haemoglobin suited for their species.

---

[16] Rudge DW. 1998. 'A Bayesian Analysis of Strategies in Evolutionary Biology', *Perspectives on Science* 6(4):341-360.

The extended evolutionary synthesis (EES)[17] expands what is recognized as causally relevant in the process of evolution. Standard evolutionary theory is gene centred and treats as evolutionary processes solely those events that change gene frequencies: mutation introduces new variants at random. Repeated occurrence of the same genetic variants is called mutation pressure. Natural selection makes adaptive variants more common through differential survival and reproduction. Genetic drift occurs through random changes in frequency of genetic variants due to sampling. In gene flow, variants enter and leave a population via migration, dispersal or mating.

The EES is organism-centred and recognizes other processes in addition to those that change gene frequencies: Developmental bias occurs when developmental processes guide organisms' forms along particular pathways. Plasticity is created when novel, potentially functional, forms are induced by the environment and subsequently stabilized by selection. In niche construction, organisms systematically modify environmental resources in ways that impose biases on descendants' development and evolution. Inclusive inheritance is the process whereby organisms inherit a wide variety of materials from their ancestors, including epigenetic marks, hormones, symbionts, learned knowledge and skills, and ecological legacies.

Evolution has frequently been characterised as 'survival of the fittest'. To an extent, this is true, if one means thereby the ability of an adaptation to be inherited and give an advantage to the organism and its descendants that means that those possessing that adaptation become dominant. Note that this adaptation applies to only one property and one group of individuals in one environmental niche.

The process of inheritance and evolution is more complex than just this gene causes this phenotype (the actual properties of the individual). Genes are not always 'turned on' and their action may be modified by other genes and the environment (internal and external). Remember that each somatic cell in the body has the same DNA yet different cells in the embryo go on to produce quite different structures.

Also, we have the strange phenomenon of apoptosis (cell suicide), where cells suicide in the interest of fitness as a whole. This is extremely important in the

---

[17] Laland K, et al. 2014. 'Supplementary information to: Does evolutionary theory need a rethink? *Nature* 514:161-164.

making of networks of neurons and other complex structures and appears to be partly self-regulated and partly monitored by other cells. Rogue cells that refuse to die may end up as cancer or viral carriers.

Despite considerable continuing debate on the mechanism(s) of evolution, one thing is for certain: evolution does occur; new species do arise.

100,000 years ago, at least six human species inhabited the earth. Today, there is just one—*Homo sapiens*. The debate over whether Neanderthals were a subspecies or a separate species seems more and more to be coming down on the side of a separate species. The evidence also suggests that it was probably competition with *Homo sapiens* that led to their extinction (though with some interbreeding).

Genetics

'A solid foundation in genetics is increasingly important for everyone.'—Anne Wojcicki
Back in the 1980s, there was considerable hype about how gene therapy—the insertion, deletion or modification of a gene or some element of a gene—could one day be used to cure, or even eliminate, some diseases. This promise has been largely unfilled for a number of reasons.

Firstly, the effect one is hoping to achieve is, by and large, confoundedly technically difficult. Secondly, there are very few diseases that can unequivocally be said to be due to the presence, absence or alteration of a particular gene—despite almost daily accounts of a gene or a number of genes being more common in a disease and therefore presumably somehow causally related to that disease. Thirdly, the interplay between various genes and between genes and the environment may lead to quite unexpected results.

The gene concept has evolved tremendously—to the extent that it is difficult to define. Our comprehension of the structure and organisation of the genetic material has increased greatly, but this has entailed the discovery of assembled genes, nested genes, overlapping genes, polyprotein genes, repeated genes, split genes, transposable genes, complex promoters, and many others.

A discussion that often gets caught up with evolution is the larger question of how much of individual and racial differences are genetic and how much are due to culture and society. Both are obviously important. Unfortunately, convictions as to their relative importance are often not evidence-based.

Environment

The ultimate aim of most conservationists is to maintain the environment exactly as it is. This is, of course, impossible. They also need to realise that there will always be species unable to cope with current conditions and therefore disappearing. And loss of a single species does not necessarily do much for the stability of an ecological system.

Perhaps the biggest environmental issue today is climate change due to global warming due to human activity, especially involving accumulation of 'greenhouse gases', mainly carbon dioxide and methane. Though a great deal of evidence has been produced to prove this proposition, the evidence has been mainly indirect and based on modelling, and there are scientists who are sceptics and have produced convincing arguments opposing the idea (e.g., Plimer[18]). As a scientist, I am surprised that no one seems to have yet attempted to ascertain experimentally the relationship between increased greenhouse gases and temperature.

Population

From 1954 onwards, there have been conferences on population sponsored by the United Nations Organisation about every 10 years. The rationale behind these conferences is largely the Malthusian concept that unchecked population growth will inevitably outpace availability of resources. The main solution proposed has been contraception, though the idea that development will both lessen the impact and eventually decrease the rate of growth has often been propounded. From my observation of various countries, the most effective way to decrease population growth may be to provide an effective social welfare system which means that parents do not have to rely on children for support. Probably because of a combination of these factors, birth rates are decreasing in the great majority of countries worldwide. However, this will not prevent a further drastic increase in population before stability is reached (if it is).

---

[18] Plimer I. 2009. *Heaven + Earth*, Connor Court Publishing, Ballan, Vic, Australia.

Another factor that cannot be dismissed is the possible influence of climate change due to global warming. While doomsday scenarios may underestimate our capacity to adapt, the adaptation may be fraught with difficulties and traumas.

Our interference with the carbon cycle with resultant effects on global temperature is only one area in which we may need to be more responsible. Another is in the nitrogen cycle, where inefficient use and overuse of nitrogenous fertilisers has changed the natural cycle.

Developed societies in particular are also guilty of enormous waste. An enormous amount of food goes to waste because it is not perfect enough for choosy consumers. In underdeveloped societies, there is also an enormous waste of food due to inefficient farming practices, poor pest control and inadequate storage facilities.

Developed countries are now quite proud of their efforts in recycling. These efforts are worthwhile, but they are also a symptom of inbuilt waste. Much of what is recycled could be reused and a large part of recycled material is due to excessive packaging. Is it really necessary for householders to remodel kitchens, bathrooms, etc. every few years? Do shopping centres really need to be completely remodelled ever few years? According to a recent colloquium at the Getty Center, the average life span of a conventionally built building (masonry and wood) is about 120 years. But for modernist buildings (reinforced concrete and glass curtain wall) it's half that: 60 years. And the replaced materials in each of these cases are more likely to go to landfill than to be recycled.

Anthropology and Sociology

Anthropology is variously defined. Perhaps the best definition is 'the scientific study of the origins of humans, how we have changed over the years, and how we relate to each other, both within our own culture and with people from other cultures' (vocabulary.com). The same source defines sociology as the study of human cultures, communities, and societies. Sociology attempts to explain how a society works.

Though both anthropology and sociology are regarded as sciences (even if 'soft sciences'), they may not always be scientific. An outstanding example is Margaret Mead's *Coming of Age in Samoa*, which was based on little but gossip, yet popularised a fundamentally misguided view about human culture. This was shown to be false by Derek Freeman in the 1980s but is still remarkably prevalent.

The Mind

There was a time when the study of the mind was the province of philosophy. Now, with the realisation that the mind is a function of the brain, the mind is as likely to be studied by physiologists, psychologists and psychiatrists as by philosophers.

Wilhelm Wundt, the first person to call himself a psychologist, created psychology as a science apart from philosophy and biology, whose task was to precisely analyse the processes of consciousness, assess the complex connections, and find the laws governing such relationships.

While psychology has to some extent proceeded along these lines, psychologists have also become involved in more 'practical' issues, such as Robert E Thayer's *The Origin of Everyday Moods: Managing Energy, Tension, and Stress*, or *Beating the Blues* by Susan Tanner and Jillian Ball.

On the other hand, there is the work of the neurologists. In the 1860s, there was a paradigm change when neurologists began to change the classification of neurological diseases from anatomically based to one of presumed causation and relationship to other states.

The two approaches sometimes merge. An example (together with physiological discoveries) is the discovery of the right and left brains and their eventual union and orchestration.

However, there are many areas of brain and mind function which are still rife with speculation with very little supporting evidence. A good example is the function of dreaming and REM sleep, for which there are many theories but scant scientific evidence. Certainly, there is no rational evidence for dream interpretation.

Psychiatry incorporates biochemistry and neuroendocrinology but falls short of being an exact science because of the difficulty of fitting such discoveries into ordinary human life. Psychiatrists tend to speak of disorders rather than diseases. A disorder is a fault in mental function produced by some defect in the brain (in the computer analogy, this may be a fault in hardware or software or firmware).

The use of imaging, especially positron emission tomography (PET), has clearly shown what happens in the brain when we think and feel, again illustrating that mind is a function of the brain.

From time to time, a controversy arises about whether intelligence is genetic or a function of environment. Originally, intelligence was defined as the inbuilt capability that allows us to reason, plan, solve problems, think abstractly, comprehend complex ideas, learn from experience, learn quickly and perform other 'higher' mental functions. Intelligence of this order applies only to humans (out of the billion or so organisms that exist or have existed). Certainly, these abilities can be developed by teaching and enriched experience, but the initial spark must be there. It is important to realise that 'there are many other mechanisms and devices than intelligence to cope with survival and reproduction, mechanisms that apparently can evolve far more easily than high intelligence'[19]

Consciousness is a term that is used in a number of different meanings and contexts. If a person loses consciousness, they completely lack awareness. However, there are many processes that occur within the brain of which we are not conscious. Thus blindsight appears to be due to the fact that only half of the vision stream is conscious. In psychology and psychiatry, the half that is not conscious (and other such processes) is said to be unconscious. Metaphysical and 'New Age' writings often use the term subconscious in this sense, whereas in psychology the subconscious is whatever we do not currently hold in focal awareness. The term subconscious is not used in psychoanalytical writings. Instead, Freud posits an unconscious mind with a will and purpose of its own that cannot be known to the conscious mind.

Altered states of consciousness differing from the normal waking state of consciousness can be induced by pharmacological, physiological or psychological means.

However we store them, our memories are quite fallible. People remember events that never happened. We also forget things that did happen, more and more as we age.

Hypnagogic consciousness occurs during the onset of sleep and may involve lucid dreams, where the dreamer is aware of dreaming, hallucinations, and other phenomena. This may relate to different types of brain waves.

---

[19] Mayr, E. 1994. 'Does it pay to acquire high intelligence?' *Perspectives in Biology and Medicine* 37(3):337-338.

Pseudopsychology

'*Psychoanalysis is that mental illness for which it regards itself as therapy.*'—Karl Kraus
Efforts to establish psychology as an exact science are hampered not only by the difficulty of studying mental phenomena but also by the amount of pseudoscience built up around it, including not only psychoanalysis but also ranging from quizzes in popular magazines to facilitated communication, enneagrams, Werner Erhard and est, evolutionary psychology, eye movement desensitisation and reprocessing, synchronicity and the collective unconscious, polygraphy, the Mozart effect, multiple personality disorder, Myers-Briggs type indicator, neurolinguistics programming, reverse speech, recovered-memory therapy, and past life regression. Many of the pseudosciences that people avidly follow, such as astrology, astrotherapy, biorhythms, cartomancy, chiromancy, fortune telling, graphology, and rumpology, depend on the Forer effect, the tendency of people to rate sets of statements as highly accurate for them personally though they could apply to many people.

Decisions

'*The more decisions you are forced to make alone, the more you are aware of your freedom to choose.*'—Thornton Wilder
'*More is lost by indecision than wrong decision. Indecision is the thief of opportunity. It will steal you blind.*'—Marcus Tullius Cicero
This impinges on the question of free will. Are we really free to choose? Our decisions are products of an intricate mix of reason, intuition and emotion. If we relied on reason alone, we'd be almost incapable of deciding anything at all.

Reason may not always be the best solution to a problem. Sometimes, simple trial and error may be the most effective and efficient approach.

Of course, you could always go back to divination using bones.

CHAPTER 3

MEDICINE

'For, medicine being a compendium of the successive and contradictory mistakes of medical practitioners, when we summon the wisest of them to our aid, the chances are that we may be relying on a scientific truth the error of which will be recognized in a few years' time. So that to believe in medicine would be the height of folly, if not to believe in it were not greater folly still, for from this mass of errors there have emerged in the course of time many truths.'—Marcel Proust
'Life is the collection of functions that resist death.'—Bichat.

Medicine has been called 'the greatest benefit to mankind' and in many ways it is so, despite many blunders. Medicine is not science, but today there is a great push towards evidence-based medicine, and medicine is becoming more and more a science and less and less an art. There is also an increasing emphasis on maintaining health rather than curing illness.

Diagnosis is more and more based on laboratory data than subjective examinations. There is a problem with this in that any change from 'normal' is often used to define a disease. An anomaly does not necessarily indicate an abnormality. Neither does a laboratory finding, such as hypercholesterolaemia, indicate a disease. Neither should a group of pathological states, such as coronary heart disease, be treated as a disease, especially epidemiologically. Finding statistical correlations between various factors and coronary heart disease and designating them as risk factors is neither logical nor scientific.

Over the course of time, pathogenic concepts have changed. We don't have to go back to ancient times when humours reigned supreme. Within the last century, we have seen many different concepts hold sway. One example is autointoxication, where bacteria in the bowel were held to cause all manner of ills (science has now shown that, unless we stuff them up with various toxins, including antibiotics, or poor diets, the bacteria in our gut perform many useful functions for us). Then there was stress as the cause of just about everything, including stomach and duodenal ulcers, which were eventually scientifically proved to be due to a bacterium *Helicobacter pylori* or sometimes ingestion of non-steroidal anti-inflammatory drugs.

Much progress in medicine has been as a result of basic research. Clinical research has also sometimes led to achievements in basic research. Many advances, however, have been due to an application of a new technology.

The search for drugs has also become more scientific, being frequently based on knowledge of the biology of a disease causation and the structure of a molecule that might interfere with its development. Any candidate drug so discovered must then be subjected to well-designed clinical trials. Sometimes, 'politics' intervenes in this process, as in the decision to allow the use of cannabis oil.

Alternative Medicine

By and large, alternative medicine is not only unscientific but also anti-scientific. This is not to say that some of the so-called natural medicines may not sometimes work—though, when they do, it is very likely to be a placebo effect. The problem is that very few of these products and theories have been subjected to scientific testing, and when they have, many have been found wanting.

A good example of a pseudoscientific approach to health is Deepak Chopra's *Ageless Body, Timeless Mind*, which wraps up a few common-sense suggestions with mysticism.

As J R King observes, 'A paradox in this scientific age is the remarkable popularity of some totally unscientific therapies', of which he instances aromatherapy.

Why is alternative medicine so popular? There are many reasons, probably the most important of which are:
- many preparations are available over the counter;
- these preparations are often seen as more natural (which in itself is a fallacy);

- you can diagnose yourself on the internet and order online;
- you've tried everything else;
- alternative medicine practitioners are often more willing to devote time to a patient and listen to a patient;
- they tap into a common desire for a more exotic and mystical experience.

For example, Tibetan medicine appeals to both the latter two factors. This is not to say that Tibetan medicine does not have some validity but rather that it has not been scientifically validated. The same could be said of Chinese medicine, although considerable effort has been made in recent time to in fact investigate some of the claims made for it. However, tests of traditional Chinese medicines confiscated in Australia from overseas travellers showed some to contain endangered species, toxins and allergens. Even in Australia, you can't always be sure what you are getting.

Some of the many forms of alternative medicine which have little, if any, scientific validation include acupuncture, alphabiotics, applied kinesiology, aromatherapy, aura therapy, ayurvedic medicine, Bach's flower therapy, chelation therapy, chiropractic, craniosacral therapy, DHEA, holistic medicine, homeopathy, hypnosis, iridology, joy touch, macrobiotics…and the list goes on. Some of these claim to be able to produce scientific studies that prove their value, but any such studies I have looked at have been very poorly conducted and failed to support the conclusions reached. One problem in evaluating these claims is that products and methods are poorly standardised between practitioners. Of course, there are those who claim that science is not equipped to evaluate these methods. I have some sympathy with this viewpoint when it relates to the difficulty of standardisation and of the isolation of various effects, but for many it seems to be more a mystic belief.

CHAPTER 4

ECONOMICS

Despite the fact that economics is becoming more and more mathematical, it is still far from an exact science. Some wit is reported to have said that the only reason economists exist is to make weather forecasters look good. This is a bit harsh, but economic forecasting tends to be about as accurate as weather forecasts. The trouble is that both rely on models using retrospective data and largely unproven theories and both suffer from a sort of 'butterfly effect'. As Paul Ormerod pointed out, building theories from the facts at the outset is more likely to yield results than pursuing abstract theories of how an ideal world should operate.

One hypothesis that seems to be well proven is that a market economy is the path to wealth for a nation, though there are many other factors.

The monetary/fiscal policy debate appears to have been settled by the experience of Japan, where reliance on fiscal measures has resulted in continued stagnation for the past 20 years.

The Clinton presidency saw the greatest period of economic growth in US history, just ahead of that under Franklin Roosevelt. Unfortunately, the policies of George W Bush reversed this. Obama again achieved strong growth. Yet, Donald Trump has succeeded in convincing enough American voters that the US is in a dire economic condition to get himself elected President. As Obama said, 'We should be guided by what works'—not by dogma.

In Australia, from very early days, economic policy has involved two pairs of opposing philosophies: protectionism v free trade; and socialism v laizzer faire. Currently, the overall trend is towards free trade and a mixed economy. During

World War II, John Curtin introduced a series of measures that increased production and ended unemployment. These measures were so successful that Australia was able to largely fund its own war effort and even to assist Britain. The succeeding Chifley government carried out a vigorous program of national development but also pursued a number of socialist objectives, including nationalising Qantas and setting up a number of government enterprises. It was brought down by an attempt to nationalise banking. Menzies' newly formed Liberal Party was committed to a mixed economy. Economic growth, high employment levels, growing overseas investment and the development of new markets led to a high level of economic prosperity. High population growth fuelled partly by measures encouraging procreation but mainly by immigration, and high government spending funded by rising taxation receipts made Australia wealthier than ever before, even allowing for periodic credit squeezes and a lack of national economic planning. Increases in tariff protection protected jobs and profits but had a negative effect on productivity and innovation. When Britain progressively abandoned Imperial Preference and moved to enter the European Economic Community, Australia's trade position deteriorated. Successive Liberal governments post-Menzies were unable to deal with this and such crises as the oil price shock, and the country (along with many others) entered a period of stagflation that saw Labor under Whitlam assume control. The failure of the Whitlam government to effectively manage the economy was a factor in its demise. The Liberal Fraser government and succeeding Liberal governments sought to regain the early post-war good times by implementing measures that worked then, but times had changed, and they failed. Economic liberalisation and deregulation of the Australian economy began with the Labor Hawke government and was carried on under Keating. Measures included progressively cutting tariffs, floating the Australian dollar, deregulating the financial system, and privatising several large government enterprises. A global recession in the early 1990s led to Labor losing power, to be replaced by John Howard, who carried on the process of economic reform and restructuring. However, WorkChoices, Howard's attempt to reform industrial relations laws, was a step too far, and a vigorous union campaign saw him replaced by the Rudd Labor government. Rudd's attempt to deal with the global financial crisis by a 'splash of cash' has been praised by some as preventing Australia sliding into recession but, even if this is admitted (which many do not, the counter-argument being that it was the strength of Australia's banks and the mining boom that saved Australia and the government expenditure only created a deficit successive governments have been unable to correct), it was very clumsily handled. His replacement by Gillard and his return to power did little to improve things and the succeeding Liberal government under revolving leaders has had little success in achieving its objectives, partly due to many of its measures being blocked in the Senate. Unfortunately, Labor doesn't seem to have any more answers. Where Australia goes from here is anyone's guess.

CHAPTER 5

ANTISCIENCE, JUNK SCIENCE, PATHOLOGICAL SCIENCE AND PSEUDOSCIENCE

'It is not reason which is the guide of life, but custom.'—David Hume.

The term 'antiscience' may embrace a number of different phenomena. In my usage, it particularly relates to attacks on endeavours in areas where science is indubitably the best, or even the only rational, approach and attempts to substitute or at least give equal time to other approaches. 'Junk science' is where there is at least some attempt to apply scientific principles, but for one reason or another, data and/or conclusions are flawed. 'Pathological science' was defined by Langmuir as research where 'people are tricked into false results ... by subjective effects, wishful thinking or threshold interactions' and includes N-rays, polywater, cold fusion, and water memory. 'Pseudoscience' includes, among many others, astrology. Other antiscientific beliefs may be just plain silly or based on superstition.

One of the more pernicious forms of Antiscience is that which has permeated academia. This has many forms, one of which is relativism, the belief that scientific findings have no absolute truth but are merely relative and subjective. At least some postmodernists have rejected scientific objectivity, the scientific method and scientific knowledge.

'Occult statistics', where computers are used to analyse masses of data to produce statistically significant but spurious correlations, used by astrologers, parapsychologists, etc. to prove the existence of the occult, are an example of 'junk science'. 'Data mining' by mainstream scientists is sometimes not too far removed from this.

Langmuir lists six characteristics of pathological science: 1. The maximum effect that is observed is produced by a causative agent of barely detectable intensity, and the magnitude of the effect is substantially independent of the cause. 2. The effect is of a magnitude that remains close to the limit of detectability, or many measurements are necessary because of the low statistical significance of the results. 3. Claims of great accuracy. 4. Fantastic theories contrary to experience. 5. Criticisms are met by ad hoc excuses thought up on the spur of the moment. 6. Ratio of supporters to critics rises up to somewhere near 50% and then falls gradually to oblivion.

Many of the absurdities in pseudoscience come about from errors in getting and interpreting evidence. Many are characterised by ad hoc hypotheses, i.e. explanations tacked on when the original theory doesn't work.

One unfortunate form of pseudoscience which has resulted in the death of children is Christian Science, which is neither Christian nor science.

Another popular (at least in parts of the USA) pseudoscience is so-called creation science, which actually has little in common with science. Unfortunately, when science is pitted against pseudoscience in the public arena, many non-scientists interpret this as establishment defending its turf against new ideas. Scientists need to show that creation science simply is not science; a process that starts with a conclusion and marshals evidence to support it while ignoring anything that does not is definitely not science.

Gaia is another pseudoscientific concept which covers a variety of opinions ranging from the fairly obvious idea that organisms have some capacity for altering the environment to suit themselves to the rather fantastic idea that the entire earth (or even the entire universe) is a single compound organism.

The field of personality and character analysis abounds with pseudoscience. Examples are phrenology, craniometry (craniology) as applied to the measurement of cranial features to classify people according to race, criminal tendencies, intelligence, etc., graphology when used to infer a person's character, metoposcopy, the Myers-Briggs Type Indicator, physiognomy, palmistry, name analysis in Kabalarian philosophy, and numerology. Dianetics, the basis of scientology, is another pseudoscience for which there is no objective experimental verification.

CHAPTER 6

THE PARANORMAL, THE OCCULT, THE SUPERNATURAL AND OTHER NON-SCIENTIFIC WAYS OF LOOKING AT THINGS

One way in which there has been progress since early history is that we are more apt to seek and accept rational explanations for phenomena than to blame these on gods or other mythological beings. However, many of us are, like Thales, the ancient Greek credited with being one of the first to seek rational explanations for many phenomena, still prone to mix rational and irrational (religious, superstitious, and other) explanations. Just look at the average women's magazine, with its plethora of irrational and easily disproven means (stars, numbers, psychics, etc.) for foretelling one's future.

Like many of us, Thales was also capable of simultaneously holding two conflicting views: not only was everything 'full of gods' but also made fundamentally of water. If there are two conflicting answers to any question, either one is wrong, or we are not asking the right question.

The Paranormal

Paranormal phenomena have been defined as those that lie beyond normal experience and scientific explanation—though attempts have been made to prove some of these 'scientifically'. This applies particularly to psi (paranormal site investigators), otherwise known as extrasensory perception (ESP), where efforts to prove the phenomenon have mainly revolved around subjects attempting to guess numbers or icons on cards. Supposedly successful experiments have involved such dodgy techniques as optional starting/optional stopping, where the only results which are kept are those that support the supposed ability.

Another example where research is characterised by cherry picking is that into near death experiences. These may be linked to out of body experiences. In either case, there appear to be adequate scientific explanations without invoking the paranormal.

Another similar example linked to ESP is clairvoyance.

Similarly, in the area of prediction, the 'Jeane Dixon Effect', the tendency to promote a few correct predictions while ignoring a large number of incorrect ones, definitely operates. The prophecies of Nostradamus are so vaguely worded they could be made to apply to whatever you wish.

An extreme example of 'research' into the paranormal occurred in China in the 1970s, when the ability of people to read messages with their ears, fingers, forehead, or some other part of the anatomy except the ears was 'tested'. Not surprisingly, nothing came of these researches.

Precognition, thoughtography, telekinesis and psychokinesis are other forms of ESP for which there is simply no valid evidence.

From 1964 to 2015, the James Randi Educational Foundation offered a prize, which eventually reached $1,000,000, to anyone who could demonstrate a paranormal or supernatural ability under agreed-upon scientific testing. Despite many attempts, no one succeeded in claiming the prize. Australian Skeptics still offers a similar $100,000 prize to anyone who can prove they have psychic or paranormal abilities. This prize also remains unclaimed. They also annually present the Bent Spoon Award 'to the perpetrator of the most preposterous piece of paranormal or pseudoscientific piffle'. The 2016 award went to Judy Wilyman, Brian Martin and the Faculty of Social Sciences at the University of Wollongong for awarding a doctorate on the basis of a PhD thesis riddled with errors, misstatements, poor and unsubstantiated 'evidence' and conspiratorial thinking; the 2017 to National Institute of Complementary Medicine and University of Western Sydney for the continued promotion of disproved and unproved alternative medicine practices; and the 2018 to Sarah Stevenson/Sarah's Day for the promotion of questionable natural health remedies via her vast network of followers.

Vitalism

Vitalism maintains that living organisms differ from non-living in possessing some property (such as a 'vital spark') or because they are governed by different principles to non-living entities. This property may be the Chinese chi, the soul, or some other poorly defined (or undefined) attribute. The existence of these properties is impossible to demonstrate, and the rational conclusion is that the mystical energies supposedly at their core—and of other such derivatives or associates as feng shui—simply don't exist. The same can be said of the prana of Hindu philosophy.

Occult and Occultism

'The fault, dear Brutus, is not in our stars, but in ourselves.'—*Julius Caesar* Act 1 Scene 3.

The occult has a powerful pull, even on otherwise rational beings. Neils Bohr is supposed to have had a horseshoe hanging on his wall even though he didn't believe it brought luck. 'But then horseshoes bring luck even if you don't believe in them', he is supposed to have said. Personally, I doubt the truth of this story, but even the most rational of us do tend to have similar quirks.

Perhaps the most popular of these, at least in western society, is astrology. Practically every newspaper and women's magazine has its 'stars' column, which innumerable people read every day and some even act on, despite the fact that it has been shown to have no scientific validity or explanatory power.

Attempts have been made to integrate astrology and various schools of psychotherapy to produce astrotherapy. This is a case where two wrongs definitely do not make a right.

Cosmobiology is a pseudoscience concerned with possible correlations between the cosmos and organic life. It purports to use scientific methods but also advocates taking account of macrocosmic and microcosmic interrelations between the cosmos and organic life that are not capable of measurement.

Michael Gauquelin's astrobiology, including the 'Mars effect', where the position of Mars at birth is said to influence athletic ability (and Jupiter to influence military ability and Venus artistic ability) has no experimental validation.

Similarly, considerable efforts have been made to prove that the moon effects humans (and other animals and plants) physiologically and psychologically by some unknown means. However, no well-conducted study has ever shown any effect. Though, for lovers, it may well be true 'what a little moonlight can do'.

The 'Bible code' (more properly, 'Torah code') has been vigorously attacked by mathematicians and others who showed glaring errors but still persists in some quarters. Cabala is another mystical method of interpreting the Bible that still has adherents. The fantasy, *The Celestine Prophecies*, also still has believers, although many of the supposed 'facts' in the plot are just plain wrong.

To show what a grip such strange beliefs may have on the public imagination, the English occultist, magician, and self-proclaimed prophet, Aleister Crowley, was voted 73rd in a 2002 BBC poll of the greatest Britons of all times (Winston Churchill was voted #1). He defined 'magick' as he practised it as 'the science and art of causing change to occur in conformity with will'. It was certainly not science and there is no valid evidence that he or any other practitioner has produced the effects they claim.

There is also no scientific evidence for the belief that crystals are conduits of mysterious forces that can be used in healing, protection and telling the future. The evidence that is advanced by occultists and New Age practitioners is entirely based on testimonials, placebo effect, wishful thinking, selective thinking, subjective validation, sympathetic magic, and communal reinforcement.

New Age

New Age beliefs involve more than crystals, but they all tend to be unable to distinguish reality from myth. Their view of truth tends to be entirely subjective: truth is whatever you want it to be. Examples are Jean Houston and the Mystery

School, Landmark Forum/Landmark Education, Silva Mind Control, tachyons and takionics.

Mystics

Mystics are people whose beliefs and ideologies relate to extraordinary experiences and states of mind. What I said above about New Age practitioners applies even more to them. In fact, many of the most characteristic beliefs of the New Age movement derive from the mystic, Edgar Cayce.

A good example is George Gurdjieff, who taught that most humans do not possess a unified mind-body consciousness and live in a state of 'waking sleep', but that it is possible to transcend to a higher state of consciousness and achieve full human potential. Neither he nor any of his followers appear to have done this.

Anthroposophy has been called the most important esoteric society in European history. It aims to develop faculties of perceptive imagination, inspiration and intuition through a form of thinking independent of sensory experience and to present results so obtained in a form susceptible to rational verification. It fails in this aim. This is not surprising given its roots in theosophy, a collection of occultist and mystical philosophies which are supposed to point the way to direct knowledge of the presumed mysteries of life and nature. There is no scientific evidence for the Akhasic records—the compendium of emotions, events and thoughts followers of these philosophies believe are encoded in a non-physical etheric plane.

The Supernatural

'The many instances of forged miracles, and prophecies, and supernatural events, which, in all ages, have either been detected by contrary evidence, or which detect themselves by their absurdity, prove sufficiently the strong propensity of mankind to the extraordinary and marvellous, and ought reasonably to begat a suspicion against all relations of this kind.'—David Hume

'There is no supernatural, there is only nature. Nature alone exists and contains all. All is. There is the part of nature that we perceive, and the part of nature that we do not perceive. ... If you abandon these facts, beware; charlatans will light upon them, also the imbecile. There is no mean: science, or ignorance. If science does not want these facts, ignorance will take them up. You

have refused to enlarge human intelligence, you augment human stupidity. When Laplace withdraws Cagliostro appears.'—Victor Hugo

The supernatural deals with phenomena not subject to the laws of nature. Scientists and other rational human beings refuse to accept that there are any such phenomena.

There are many supposed supernatural beings—angels, fairies, ghosts, houris, jinn, etc. There is no evidence worth considering for the existence of any of these.

For all that, people still pay to visit supposedly haunted houses, and spiritualism in one form or another is still big business.

ETs, UFOs and Other Strange Phenomena

There is no solid evidence that extra-terrestrials have ever visited Earth or that UFOs are flying saucers come from somewhere out there or for alien abductions. Erich von Daniken's evidence' for ancient astronauts is weak, and Zecharia Sitchins' theory of human origins involving astronauts is purely fanciful. All the evidence points to crop circles being made by human hand. The face on Mars turned out to be just another mesa on closer examination. Maarfa lights and many other lights sometimes claimed to be UFOs or paranormal are all subject to natural explanations.

Then there is the completely fantastic hollow earth theory, which is easily disproved by basic scientific observations.

Lawrence David Kusche concluded that the legend of the Bermuda Triangle is a manufactured mystery, perpetuated by writers who either purposely or unknowingly made use of misconceptions, faulty reasoning, and sensationalism[20]. The same can, perhaps, be said of the Loch Ness monster, sightings of which appear to be hoaxes, wishful thinking, or the misidentification of mundane objects. There may also be similar explanations for sightings of Big Foot, bunyips, sasquatch, yeti and yowie. I don't know of any reported sightings of unicorns, vampires or werewolves in recent times, but if there were any, I'd expect that the explanation would be similar—or just a completely delusional observer. As regards

---

[20] *The Bermuda Triangle Mystery: Solved* (1975)

vampires and werewolves, modern medicine may have provided the answer in the form of rabies, but this is probably impossible to verify without completely unethical experiments.

Why do people still believe in all these things? Recent psychological research indicates that belief in conspiracy theories, the paranormal and pseudoscience all tend to go together and to depend on a preponderance of intuitive over reflective thinking.

There are still people who believe the Apollo moon landings were faked, despite clear evidence for the landings and against the spurious arguments advanced to 'prove' they were faked. On the other hand, all kinds of bizarre urban myths surface and are widely believed at least for some time. The internet has made it easier both to propagate and to debunk such myths and legends.

Fraud and Hoaxes

Fraud, the deliberate fabrication or falsification of evidence with the intent to deceive or mislead, does occur in science, as in other human endeavours. Piltdown man is one of the most famous frauds in the history of science. The amazing thing is that it took over 40 years to show that it was a fraud—perhaps not so surprising when you consider the state of human anthropology at the time.

Fraud is, however, much more common in those who deal in the paranormal. This includes the many who have claimed to 'channel' various spiritual entities, such as Ramtha, and gurus, such as Sai Baba, who claim to be reincarnations of God or a saint or other worthy.

Stepping down a level or two, Uri Geller's psychic powers are indubitably nothing but stage magic.

Some frauds are just lies made to look plausible enough to convince people who want to believe they are true, such as the *Protocols of the Elders of Zion*, a totally fraudulent document used by the Nazis and others to promote anti-Semitism.

On a more mundane level, we have chain letters, Ponzi schemes and pyramid schemes (the latter sometimes dressed up as multi-level marketing). These all depend on having an infinite chain of suckers.

Propagandists are experts in creating and spreading false reports which, especially in time of war, people are prepared to believe of their enemies no matter how bizarre or unlikely they may appear. Good examples from World War 1 are the Kaiser's supposed telegram to his commander telling him to crush the 'contemptible little army' of the British, the angels of Mons, crucified soldiers, and the German corpse reprocessing program (this latter revived in the Second World War).

CHAPTER 7

RELIGION

'When the Church says that, in the dogmas of religion, reason is totally incompetent and blind, and its use to be reprehended, this really attests the fact that these dogmas are allegorical in their nature, and are not to be judged by the standard which reason, taking all things sensu proprio, can alone apply. Now the absurdities of a dogma are just the mark and sign of what is allegorical and mythical in it.'—Schopenhauer

'Monotheistic religions alone furnish the spectacle of religious wars, religious persecutions, heretical tribunals, that breaking of idols and destruction of images of the gods, that razing of Indian temples and Egyptian colossi, which had looked on the sun 3,000 years: just because a jealous god had said, "Thou shalt make no graven image".'—Schopenhauer

'Religion is a subcategory of supernaturalism that was formulated during the medieval period with the spurious and dangerous quest to link supernaturalism with scientific knowledge, and this quest has continued.'—Jacob Pandian.

Religion could easily fit into either of the two preceding chapters but probably deserves one of its own.

There are those who believe that religion and science are compatible. We are at the stage of history in which Christianity is attempting to look as though it endorses science, as in Pope Francis' declaration that evolution and the Big Bang are real and not incompatible with the existence of a creator, arguing instead that they require it. This is nonsense: the two modes of thought are diametrically opposite.

The case for atheism and against God has been well put by, e.g., George H Smith.

'It is not my purpose to convert people to atheism... (but to) demonstrate that the belief in God is irrational to the point of absurdity. If a person wishes to continue believing in a god, that is his prerogative, but he can no longer excuse his belief in the name of reason and moral necessity[21].'

More recently, Richard Dawkins has pretty well destroyed the God hypothesis in his book *The God Delusion*. Eighty years before that, Bertrand Russell did the same. He also gave good reasons not to be a Christian, including showing that Jesus was not as wise as claimed nor as moral (though many of his better moral principles are not obeyed by his followers).[22]

Very few supposed Christians have actually read the Bible. If they read it from cover to cover, they firstly will be struck by the many inconsistencies (which is probably not surprising in a multi-author work, but it is supposed to be the word of God). Even more so, they will be appalled by the litany of crimes against humanity supposedly carried out by the Israelites at the behest of Jehovah. 'That's the Old Testament, before Christ,' you might object, but I fail to find anything moral in a God who would make a creature designed to sin and then condemn him to eternal torture when he does. One can only agree with Peter Sellers, who after completing reading the Bible through because it was the only book available to him, concluded, 'God is a bastard!'.

The Koran is certainly no better. Schopenhauer summed it up nicely:

'Consider the Koran... this wretched book was sufficient to start a world-religion, to satisfy the metaphysical need of countless millions for twelve hundred years, to become the basis of their morality and of a remarkable contempt for death, and also to inspire them to bloody wars and the most extensive conquests. In this book we find the saddest and poorest form of theism. Much may be lost in translation, but I have not been able to discover in it one single idea of value.'

What of the Eastern religions? Buddhism is, at least in principle, considerably more humane. Buddhism believes in reincarnation. One achieves nirvana by becoming selfless, thereby eliminating desire and consequent suffering. However, if you eliminate desire, you either die or become a kind of robot or zombie.

Reincarnation is an ancient concept. There are several versions differing in detail, but a common thread is a number of rebirths until one has learnt enough to be fit

---

[21] George H Smith. 1974. *Atheism—the Case Against God*. Nash Publishing.

[22] Bertrand Russell, *Why I Am Not a Christian*. 1927. Watts & Co.

for nirvana. Though superficially attractive, it suffers from the irremediable defect that it would result in a diminishing supply of souls rather than the ever-increasing supply necessary for a burgeoning population. That Pythagoras, one of the first mathematicians of note, believed in it (and in his 'music of the spheres') shows again how humans can entertain a mix of rational and irrational beliefs.

The 'Golden Rule', 'Do as you would be done by', is certainly a good basis for a harmonious society. Virtually every society from ancient Egypt to the present day has had some form of this rule, though it has not always been well followed. Confucius espoused a version and emphasised the concept of public service but he also introduced the concept of 'everyone to his or her station in life'. Hinduism, with its caste system, is an extreme example of the cruelty which this can produce.

The Taoist 'doing without doing' is an example of a paradox resulting from a misstatement. What is implied here is unselfconscious thought or action. The thought or action is supposed to come from the universe without self-conscious intervention. I would deny this, instead crediting the subconscious. However, I am prepared to accept Lao Tzu's contention that some things cannot be communicated but must be experienced. Also, it must be admitted that Sun Tzu's *The Art of War* was a practical application of Taoist principles.

If a belief in religion is so irrational, why do so many people persist in this belief? Religion does fulfil certain needs for certain people, including an explanation of sorts of creation, and a promise that there is some meaning to life and hope for life after death. In Voltaire's words, *'If God did not exist, it would be necessary to invent him'*. In fact, any serious scientific investigation would soon conclude that man did in fact create God, rather than God creating man. There may well be what Jesse Bering has called 'the God instinct', though 'instinct' is definitely a misapplied term, since religious beliefs are, for the vast majority of people, learned. Also, many professed believers are, at heart, not really believers but go along with the forms of belief because of community pressures.

However, it is apparent that belief in God is not likely to quickly disappear, even among scientists. In 1916, a survey of 1000 scientists found 42% believed in God; in 1986, this figure had only dropped to 39%. However, among members of the National Academy of Scientists, only 10% express some form of religious belief.

Phillip Petersen

CHAPTER 8

ETHICS

Religion is often presented as the source of ethics and morals. This argument is hard to maintain in the face of the unethical, cruel, immoral behaviour besetting many religions. In fact, at least for Judaism, Christianity and Islam, the evidence is more the other way.

For all that, the world wouldn't be too bad a place if we all practised the Christian virtues of prudence (acting appropriately at any given time), justice or fairness, temperance (restraint, self-control, moderation and abstention), and courage. One could easily dispense with faith (belief in God) and the hope of eternal life (or making sure you go to heaven and not hell) and still be ethical. Neither does being charitable really require a belief in God.

Science and the scientific method also have little to contribute to arguments about ethics, even though successful science depends on scientists being ethical. This is not to say that bad science may not still be ethical. Cold fusion may be pathological science but (unless there is actual fraud) it is not unethical.

A big question on which science can give little help is abortion. The Christian Church has long condemned abortion (and still does), mainly on the basis of the commandment 'Thou shall not kill'. But this commandment has always been hedged with exceptions. The Israelites who received it went on to commit genocide, according to the Bible on Jehovah's orders. Despite Christ's precept of turning the other cheek, the right to self-defence is almost universally acknowledged. It can justifiably be argued that this includes cases where a mother's health is endangered by a pregnancy. Few systems of ethics argue that it is incumbent upon us to sacrifice our lives for others—although it may sometimes be laudable to do so—but there are those who believe that the rights of the unborn child should supersede those of the mother. The main argument against abortion revolves, of course, around the right to life of the child. Is this (or should it be) an absolute right? How is it moral to insist on the birth of a brain-damaged child who will suffer a miserable existence until eventually dying a horrible death and in the meantime putting the parents through a living hell? On the other hand, aborting a foetus because it happens to be female would be thought by many to be immoral and unethical. Abortion raises a number of moral questions, many of which are not susceptible to logical analysis.

Unfortunately, medicine currently shows an intense preoccupation with preserving life at all costs, regardless of the fact that the existence they are preserving is not one they would choose for themselves. This may sometimes be based on religious beliefs or fear of legal and other repercussions but may also be an inability to admit failure. In modern times, turning off life support in helpless cases has become more and more accepted.

The ethics of suicide, assisted suicide and euthanasia is hotly debated but remains murky. Although the concept of suicide as a sin is no longer as widely or as strongly held, suicide is still illegal in a number of countries. Assisted suicide and euthanasia are illegal in most jurisdictions. Despite this, many people meet their deaths as a result of high doses of painkillers and/or withdrawal of food and drink, a process that is not regarded as illegal even though it obviously hastens death.

I maintain that one has the right to exercise some control over the time and manner of one's death. Norman Cousins claimed that 'Death is not the ultimate tragedy of life. The ultimate tragedy is depersonalization'. For Van Rensselaer Potter, the time for him to die with personhood would be when his family, personal physician, and social worker agreed that there was no other choice than to place him in a nursing home. Others will no doubt have a different cut-off point.

It seems to me somewhat anomalous that it is almost universally regarded as the right thing to do to humanely end an animal's suffering but to deny this privilege to a human. Of course, this apparent regard for the animal may arise more from considerations of one's own convenience and welfare. However, the old Judaeo-Christian view that all other creatures were created solely for humans benefit is slowly disappearing and Schopenhauer's view that 'boundless compassion for all living beings is the surest and most certain guarantee of pure moral conduct' slowly taking over.

Ethics today is a growth industry. Professional ethicists are called upon to pronounce on the ethicality of such things as cloning and stem cell research. Should a scientist take note of these pronouncements, especially as they relate to her or his research? The relationship between ethics and science can be a prickly one, but neither can afford to ignore the other. Ethical considerations can certainly prohibit some scientific experiments, and ethics needs to take into account what science tells us about human nature, the neurological basis of morality and the evolution of morality.

A fairly recent development in ethics is global bioethics. 'Global bioethics in this sense is the new language of humanitarianism, emphasizing that we are citizens of the world who have responsibilities to each other. Distance and borders are morally irrelevant'[23]. The goal of global bioethics is to attempt to develop a morality that places long-range goals of human survival ahead of short-term economic gains that biological and cultural evolution have made pre-emptive. Science does have a role to play in this. Global bioethics could well serve as the credo of the non-Christian scientist—perhaps combined with the atheistic materialism of Paul Henri Thiry.

---

[23] Have, Hank Ten; 2016; Global Bioethics: An Introduction. *Published by Routledge (Taylor & Francis)*

CHAPTER 9

PHILOSOPHY

'It has been said that man is a rational animal. All my life I have been searching for evidence which could support this.'—Bertrand Russell

There are several conceptual puzzles in biology which can be related to the specific findings of the biological sciences only by complex chains of argument which are traditionally the province of philosophy. The rise of scientific philosophy in the late nineteenth century was accompanied by a change in the conception of both science and philosophy. The central questions of the philosophy of science are what qualifies as science, the reliability of science, and the ultimate purpose of science.

Science and philosophy may not always concur. A number of 'scientific' definitions of life and death compete with philosophical ones—and possibly also with 'common sense' ones. The same could be said of definitions of 'self', even the self/nonself discrimination that is the basis of immunology. The view of mind and body as distinct is not today that of many scientists, where the mind is felt to be more of a process of the brain and perhaps other body structures.

However, the great majority of scientists accept naturalism, the philosophical belief that everything arises from natural properties and supernatural and spiritual causes do not exist.

Metaphysics attempts to answer the two questions: what exists? and what is the existing thing like?

Ontology is a branch of metaphysics dealing with being. Its methods are highly hypothetical but can be adapted for various technological applications in information science and in engineering.

Logical positivism sought to legitimise philosophical discourse by putting it on a basis shared with the empirical sciences. Its central thesis is that only statements verifiable through empirical observation are cognitively meaningful. This is rather in accord to my view of the sensible universe, but it needs to be remembered, as Karl Popper pointed out, that, to be truly scientific, an assertion must be capable of disproof. I might here be accused of scientism. So be it. I firmly believe that the only way we can apprehend reality is through our senses and the only way we can establish the validity of our observations and their ascendants and consequences is through logic and rationality, the foremost form of which is the scientific method.

Goethe was scientific in his willingness to test his convictions against new evidence.

The argument between determinism and free will has been a long-lasting one. Popper maintained that there are no scientific reasons in favour of determinism[24]. I believe that we all have will, but it is never entirely free. In other words, our behaviour will always have proximate and remote causes that emanate from a huge range of factors—to the extent that they largely defy scientific analysis.

A branch of philosophy which largely defies philosophical and scientific explanation is aesthetics, which deals with questions of beauty and artistic taste. The American Society for Aesthetics welcomes anyone interested in the philosophy of art, art theory or art criticism, but surely that is not all there is to aesthetics.

---

[24] Popper K, 1950, 'Indeterminism in Quantum Physics and in Classical Physics', *Brit. Journ. Philos. of Science*, 7

CHAPTER 10

CULTURE AND THE HUMANITIES

Culture is a term that is used with a variety of meanings:
- the 'finer things of life'—essentially the humanities, especially art and 'classical' music;
- a softer sort of existence as opposed to barbarism;
- the sum total of behavioural beliefs and practices of a group or nation.

Multiculturalism is the acceptance of a number of different cultural mores within a country—as long as they don't conflict with certain basic tenets that define the larger culture of the country. For example, Sharia law is not permitted to override the established law of the country; 'female circumcision' is not permitted; forced marriage, especially of underage girls, is banned; religious practices involving cruelty to humans or beasts are proscribed; etc.

Osler firmly believed that the education of a physician should combine science and the humanities. The humanities use methods that are primarily critical or speculative and have a significant historical element—as distinguished from the mainly empirical approaches of the natural sciences.[25]

These days, just about everything is a 'classic', but in the humanities, studying the classics means delving into the cultures and achievements of the ancient Greeks

---

[25] Humanity" 2.b, *Oxford English Dictionary* 3rd Ed. (2003)

and Romans. The remarkable thing about such studies is that they reveal how people thousands of years ago pondered the same questions that we do, sometimes reaching the same conclusions and sometimes different and sometimes merely ending up as confused as we ourselves are. We should also learn the salutary lesson of how far wrong the most accomplished, most intellectual, and most notable thinker can be.

History and Pseudohistory

'Philosophy of science without history of science is empty; history of science without philosophy of science is blind.'—Immanuel Kant.

The historical method consists of the examination of primary sources and other evidence in an attempt to establish an internally and externally consistent account. As in science, new discoveries may challenge the established account and lead to revisions.

Pseudohistory purports to use the same methods but is intellectually inconsistent with the historical record and common-sense understanding and is based upon or derived from a theory which the work is intended to establish. There are numerous examples, from von Daniken's spacemen to Afrocentric ideas to *The Holy Blood and the Holy Grail*.

As we delve into history, we are apt to discover that, despite all our modern inventions that have radically changed our world, there truly is nothing new under the sun and humans are not basically different from what they were as far back as we can go.

In following history, one also learns the truth of the words of WB Yeats, 'The history of a nation is not in parliaments and battlefields, but in what people say to each other on fair days and high days, and in how they farm and quarrel and go on pilgrimage.'

Geography

'Just as all phenomena exist in time and thus have a history, they also exist in space and have a geography.'— United States National Research Council, 1997.

Geography is a science devoted to a systematic study of the Earth and its features. The interaction of geography with humanistic studies, such as culture, economics, politics, all other forms of human endeavour, is also important. A glance at the contents of *National Geographic* magazine will give an idea of the range of human endeavours covered.

Language

Languages may be studied scientifically or as part of the humanities. The scientific study of language is known as linguistics. Language has also been a subject of much debate and study in philosophy. Of course, the study of languages to be able to converse in them and the study of their use in literature are central to the humanities.

Linguistic and archaeological research indicate that all the European languages and a number of Asian languages derive ultimately from the Proto-Indo-European language, which originated somewhere east of the Dnieper River on the grasslands of the Eurasian steppes—an area from which archaeology and mythology indicate a large number of the tribes of Europe came.

Naturally, linguistics has attracted its share of pseudoscience, such as reverse speech and its cousin, backwards satanic messages. Glossolalia (speaking in tongues) is another phenomenon where nonsense is manipulated to supposedly represent sense. Speed reading has some basis in fact, but greatly exaggerated claims ignore physiological and psychological facts.

Those who claim that English is the universal language have not travelled very far. Even in countries like France and Germany, if your only language is English, you will more often than not have difficulty finding someone to give you elementary directions let alone hold a meaningful conversation with. With the advent of translation apps and programs, this will probably only increase.

Politics

'Whoever could make two ears of corn, or two blades of grass, to grow upon a spot of ground where only one grew before, would deserve better of mankind, and do more essential service to his country, than the whole race of politicians put together.' — Jonathan Swift, *Gulliver's Travels*.

Politics is difficult to study scientifically. As Lawrence Lowell put it, *We are limited by the impossibility of experiment. Politics is an observational, not an experimental science.'* Nonetheless, experimental research has become a growing area of political science.

Scientific assumptions unavoidably inform political theory. Political science seeks to base society upon reliable and replicable scientific laws of human behaviour. Political theories all tend to flow from basic facts of human nature. That is not, however, all there is to it. Politics is concerned with knowing the facts of human nature so as to be able to utilise and manipulate (even alter) behaviour.

Science is often said to be apolitical, but science and politics are plainly related: science is the pursuit of knowledge, knowledge is power, and power is politics. Science provides us with facts; what we do with those facts is deeply political.

Politics and science often do not coexist happily. Science is frequently manipulated for political gain. Perhaps the best-known examples of this are Lysenkoism in Russia and eugenics under the Nazis, but it is a constant concern in all countries.

The Arts

Although we don't need to know anything about the science of art, whether performing or visual, to appreciate it, both academic researchers and actual practitioners subject art in all its forms to scientific, as well as critical, analysis.

CHAPTER 11

EDUCATION

Since the beginning of human life, education has had two main objectives: ensuring the survival of the individual; and preserving the society. These remain the basic objectives of education even today—though, in our more secure times, they are more likely to be expressed in terms of happiness or achievement than in terms of sheer survival. Of course, there has always been (and remains) considerable tension between the needs of the individual and the needs of society.

Throughout most of recorded history, education beyond sheer survival skills was aimed at producing the cultured gentleman. With the industrial revolution and increasingly sophisticated methods of production and commerce, education of the masses in the basic skills of reading, writing and arithmetic became more and more prominent. The advent of true democracy (contrary to popular representations, ancient Greece was not truly a democracy but rather an elitist society founded on a broad base of slavery) saw an ever-increasing push for a general public capable of making informed decisions about who should rule them, with a consequent broadening of the aims of education.

One of the continuing tensions between competing educational systems has been the emphasis placed on content or process, with gross excesses in claims on both sides of the debate. This largely relates to differing philosophical views which, on the one hand, see children as empty vessels to be filled and, on the other, as individuals who, by some mystical Platonic process, already contain all knowledge and only require it to be discovered.

'Traditional' Teaching

'Traditional' teaching is characterised as teaching in which content predominates and where the authority of the teacher and the text book is dominant. Opponents object to it not only because they claim it does not encourage critical thinking but because they see it as enforcing establishment power structures—in contrast to their systems, which are claimed to empower students and the citizens they later become.

One of the chief objections to 'traditional' learning has been its emphasis on rote learning. Designers of competing systems often go to extreme lengths to avoid any semblance of this. What they apparently fail to realise is that rote learning is the only possible way of learning such things as the bases of language and mathematics, since these are fundamentally true only be definition and not through any process of logic.

In material on Education Queensland's website, there is a very biased comparison of 'traditional' teaching versus 'authentic' teaching, in which they claim that 'traditional' methods are based on Skinner's rat experiments. This is highly emotive nonsense, but it is true that the experimental analysis of behaviour has led to an effective technology applicable to education, which will be more effective when it is not competing with practices that have had the unwarranted support of mentalistic theories.

'Progressive Education'

The term 'progressive education' has been used to describe ideas and practices that aim to make schools more effective agencies of a democratic society. This is to be achieved by cultivating a respect for diversity and by the development of critical, socially engaged intelligence. Such systems draw heavily on the philosophies of Plato, Rousseau and Dewey.

In my opinion, these philosophers have expected both too much and the wrong things from education. It is fine for education to touch the soul, but it is a moot point if it should turn it inward, as both Plato and Rousseau suggest. Education should involve the formation of habits of behaviour and learning but these are not, despite Rousseau, natural habits. Despite what all three philosophers would have us believe, even the best moral education cannot guarantee an improved society.

More than this, these philosophies are harmful if they direct education away from what education can and ought to do: introduce the young to what their elders believe is the best that has been done in the various forms of knowledge and experience that have been developed. A society which makes no distinction between education and training for problem solving is more barbaric than any tyranny.

The Australian education system is currently thoroughly imbued with these ideas.

Postmodernism

Dewey does, however, diverge from Plato in one important respect: he could well be designated as the first postmodernist, though he was to an extent foreshadowed by Hegel and Nietzsche. One of the central tenets of postmodernism is that reality and knowledge are neither eternal nor universal but change over time and vary from community to community.

Foucault's overly conspiratorial view is that knowledge and power cannot be separated, since knowledge embodies the values of those who are powerful enough to create and disseminate it.

Rorty asserts that the interaction between teacher and taught should be a dialogue or conversation, in which there is mutual influence rather than simple transmission from one to the other. The difficulty with this is that many followers seem to think that structure and content are unnecessary. The point about a democratic approach is not that structure and content are unnecessary but that students should have a major say in how their learning is structured and what content is made available to them. The only problem with this is that it implies students have foreknowledge of what knowledge they need—an often dubious proposition.

Unusually for postmodernists, Rorty also maintains that *'lower education is mostly a matter of socialization, of trying to inculcate a sense of citizenship. (It) should aim primarily at communicating enough of what is held to be true by the society to which the children belong so that they can function as citizens of that society. Whether it is true or not is none of the educator's business, in his or her professional capacity'*. This is the point where most postmodernist educators revert to the 'progressive education' standpoint, that it is their job to produce ideal citizens for the ideal world they will produce.

Because of the nature of postmodernism, many of the arguments of its advocates in education tend to fall into either of two categories of uselessness:

They display an easy pragmatism which, while claiming to be open and tolerant, is merely superficial, since it fails to develop and use theory; its doctrines thus become dogmatic assertions, without explanation or justification.

They regard postmodernism as a theoretical key that unlocks practice, forgetting that theory must be fundamentally rooted in practical experience if it is to be of value.

Outcome-Based Education

In contrast to a content and time-based method, outcome-based education (OBE) specifies the outcomes students should be able to demonstrate on leaving the system. These outcomes are derived from a negotiated vision of the skills and knowledge students need to be effective in whatever they have been learning. OBE focuses on the end-product and defines what the learner is accountable for, rather than telling teachers how to teach or students how to learn or specifying the steps in between. Learning outcomes determine what is taught and assessed and can help to identify what is and is not essential.

At the university level, it is well established that outcome-based learning improves learning outcomes. Well written outcomes can be very useful guides for curriculum design and assessment and can be extremely helpful to students when attempting to engage in self-directed learning.

An outstanding example of the methods that should be used in developing objectives and learning outcomes and relating them to assessment is provided by guides produced for internal use by the Centre for the Advancement of Teaching and Learning at the University of Western Australia. Objectives are clearly defined, observable, measurable and valid. Learning outcomes clearly indicate the conditions, standards and terms under which objectives are met and involve knowledge, comprehension, application, analysis, synthesis and evaluation.

When, however, it comes to lower (primary and secondary) education, the position seems to be quite different. Aims, objectives and outcomes are typically written in a manner so general that they do not guide teaching or assessment, or in a way that is not measurable. Much of this relates to the fact that desired outcomes are largely attitudinal or concerned with process rather than product.

A good example of the meaningless and useless material produced by State education departments is this gem from Education Queensland:

The overall learning outcomes for the common curriculum in the P to 10 years are defined in terms of:

Learnings expected as a result of schooling experiences through the P to 10 years of the common curriculum that may assist students to become lifelong learners, achieve their potential and to play an active role in their family and work life.

Elements that are common to all learning areas organised and presented using the following valued attributes of a lifelong learner:

Knowledgeable person with deep understanding
Complex thinker
Creative person
Active investigator
Effective communicator
Participant in an interdependent world
Reflective and self-directed learner.'

Perhaps an even better example occurs when we get to talking about 'core content':

The core content is derived directly from the core learning outcomes, which are the primary tools for planning learning experiences and assessment tasks. Students will engage with the core content when they are provided with opportunities to demonstrate the core learning outcomes.'

If this is the standard of communication skill (not to mention the capacity for clear thinking) the graduates of these new curricula are going to achieve, God help us!

Not surprisingly, many teachers flounder when trying to apply these guidelines. All the State Education Departments do provide in-service training. Unfortunately, most of these are similar to the written material in being strong on theory and very short on practice. Education Queensland recently boasted that some 30% of its teachers have attended some sort of in-service training—not really a high percentage at the best. And then you discover that this figure includes everyone who turned up for even an hour's presentation of a paper telling teachers to become more political

Though there is no real reason for it to be so, OBE in Australia at the lower level has followed the American lead and placed an overwhelming emphasis on each student's social, ethical and emotional development, at the expense of cognitive skills. This is, of course, following the agenda of 'progressive education'. Similarly, supposed training in higher order thinking skills are largely used to introduce postmodernist concepts of the absence of absolutes in any beliefs (except, curiously, those of other cultures and religions, which are often regarded as not to be questioned).

Again, while there is no good theoretical reason for the close linkage, collaborative learning has become an integral part of OBE in most versions, and certainly in lower education in Australia. In one study, university students involved in team learning identified a strong sense of teamwork and the development of interpersonal skills and autonomy in their learning, while supervisors reported that the students had a greater ability to solve problems and to take responsibility for their learning. However, a recent review found very few methodologically sound studies on group-based learning, while a number of studies have shown problems for students (mainly conflicts and stress) and for teachers (frustration at the lack of participation by some students and the fact that high achieving students did most of the work).

The latest bit of 'edubabble' to spread through Australian lower education (as a part of OBE) is 'productive pedagogy'. As an example, Education Queensland lists 22 productive pedagogies:

Intellectual quality.
Higher-order thinking.
Deep knowledge.
Deep understanding.
Substantive conversation.
Knowledge as problematic.
Metalanguage.
Connectedness.
Knowledge integration.
Background knowledge.
Connectedness to the world.
Problem-based curriculum.
Supportive classroom environment.
Student direction.
Social support.
Academic engagement

Explicit quality performance criteria.
Self-regulation.
Recognition of difference.
Cultural knowledges.
Inclusivity
Narrative.

All these terms are so poorly defined as to be almost meaningless and schools are left to make what they will of them. What they usually do with them is dutifully put them in their curriculum plans, as in these extracts from one such:

'When students of all backgrounds and ability levels are expected to undertake work of high intellectual quality, overall academic performance improves and equity gaps diminish. Classroom practices that engage students in the solving a particular problem of significance and relevance to their worlds—be it community, school based or regional problem—provide the greatest opportunity for connectedness to the world beyond the classroom. Strategies that promote supportive social environments have high expectations for all students, make explicit what is required for success, and foster high levels of student ownership and motivation.'

This is all sheer rhetoric and full of untested (and largely untestable) claims.

CHAPTER 12

THE MEANING OF LIFE

'We need not only a purpose in life to give meaning to our existence but also something to give meaning to our suffering. We need as much something to suffer for as something to live for.'—Eric Hoffer.

'Passionate hatred can give meaning and purpose to an empty life.'—Eric Hoffer.

'A passionate obsession with the outside world or the private lives of others is an attempt to compensate for a lack of meaning in one's own life.'—Eric Hoffer.

'Life has not been done to understand it but to live it.'—George Santayana.

'Music is essentially useless, as is life.'—George Santayana.

'Meaning is not something you stumble across, like the answer to a riddle or the prize in a treasure hunt. Meaning is something you build into your life. You build it out of your own past, out of your affections and loyalties, out of the experience of humankind as it is passed on to you, out of your own talent and understanding, out of the things you believe in, out of the things and people you love, out of the values for which you are willing to sacrifice something. The ingredients are there. You are the only one who can put them together into that unique pattern that will be your life. Let it be a life that has dignity and meaning for you. If it does, then the particular balance of success or failure is of less account.'—John Gardner, Personal Renewal, http://www.pbs.org/johngardner/sections/writings_speech_1.html

'What is the meaning of human life, or of organic life altogether? To answer this question at all implies a religion. Is there any sense then, you ask, in putting it? I answer, the man who regards his own life and that of his fellow-creatures as meaningless is not merely unfortunate but almost disqualified for life.'—Albert Einstein, *The World as I See It*.

'Through all ages men have tried to fathom the meaning of life. They have realized that if some direction or meaning could be given to our actions, great human forces would be unleashed. So, very many answers must have been given to the question of the meaning of it all. But they have been of all different sorts, and the proponents of one answer have looked with horror at the actions of the believers in another. Horror, because from a disagreeing point of view all the great potentialities of the race were being channelled into a false and confining blind alley. In fact, it is from the history of the enormous monstrosities created by false belief that philosophers have realized the apparently infinite and wondrous capacities of human beings. The dream is to find the open channel. What, then, is the meaning of it all? What can we say to dispel the mystery of existence? If we take everything into account, not only what the ancients knew, but all of what we know today that they didn't know, then I think that we must frankly admit that we do not know. But, in admitting this, we have probably found the open channel.' —Richard Feynman, *The Pleasure of Finding Things Out: The best Short Works of Richard P Feynman.*

The question, What is the meaning of life?, does not have an easy answer, firstly because it may mean different things.

If you mean to enquire why life exists, the shortest and probably most accurate answer is that it is one of those things that, given the right conditions, just happens.

Of course, religionists of any persuasion deny this and claim that God (whatever form he may take in the particular religion) created life the same way he created everything else. I have gone to some length in the chapter on religion to express my reasons for not accepting this explanation. One of the reasons for humans to create a creator of the universe is to give meaning to life. This doesn't really work, since no reasonable explanation for why God should create humans is forthcoming. I find the claim made by adherents of some religions that God made humans mainly, if not solely, to praise him rather nauseating. I guess an omnipotent God is entitled to a good dose of ego, but...

Several religions view our life on Earth as a kind of educational and testing experience, where we may learn and develop to the extent that we are judged worthy to join God in some form of heaven (where the chief occupation again seems to be flattering God's ego by praising him, at least in some versions) or to actually become part of God, as in some Eastern religions.

The strange thing is that *Ecclesiastes* seems to take the view (which I am inclined to share) that there is no inherent meaning to life and the best that we can do is to enjoy the simple pleasures that life can bring while we are alive.

What many mean when they search for the meaning of life is not whatever cosmic significance there may be to life but how the individual can invest her or his life with meaning. This, of course, varies with the individual (and so it must to be at all meaningful). A commonly expressed (but rarely observed) goal is to help as many people as possible. The trouble with this is that, if we all made this our raison d'etre, we would need, as Murdoch observed in one of his essays, to form a society to be done good to.

To some, family and home is sufficient reason for living. For others, it is work or passing on their knowledge in one way or another.

What most people fear is not death (though we would prefer it to be quick and painless rather than the drawn-out miserable affair now demanded in most 'advanced' societies) but non-existence. This fear may be somewhat assuaged by thoughts of surviving children and/or works left behind.

CHAPTER 13

CONCLUSION(S)

'I have never met a man so ignorant that I couldn't learn something from him.'—Galileo Galilei
'All truths are easy to understand once they are discovered; the point is to discover them.'—Galileo Galilei
'It vexes me when they would constrain science by the authority of the Scriptures, and yet do not consider themselves bound to answer reason and experiment.'—Galileo Galilei
'By denying scientific principles, one may maintain any paradox.'—Galileo Galilei
'I think that in the discussion of natural problems we ought to begin not with the Scriptures, but with experiments, and demonstrations.'—Galileo Galilei
'In questions of science, the authority of a thousand is not worth the humble reasoning of a single individual.'—Galileo Galilei
'The good life is one inspired by love and guided by knowledge.'—Bertrand Russell
'I would never die for my beliefs because I might be wrong.'—Bertrand Russell
'Fear is the main source of superstition, and one of the main sources of cruelty. To conquer fear is the beginning of wisdom.'—Bertrand Russell
'Whenever a theory appears to you as the only possible one, take this as a sign that you have neither understood the theory nor the problem it was intended to solve.'—Karl Popper
'Science may be described as the art of systematic over-simplification.'—Karl Popper
'In so far as a scientific statement speaks about reality, it must be falsifiable; and in so far as it is not falsifiable, it does not speak about reality.'—Karl Popper
'Science must begin with myths, and with the criticism of myths.'—Karl Popper
'Good tests kill flawed theories; we remain alive to guess again!'—Karl Popper

'No rational argument will have a rational effect on a man who does not want to adopt a rational attitude.'—Karl Popper

All we can really know about anything in the universe is what we can obtain through our senses. This includes information we receive through sight or sound from others. As we all know, our senses are not infallible, and the information received must always be tested. This is even more true of information received from others. The only real path to knowledge is verified experience.

Being able to decide between the known, the unknown and the unknowable may well be the beginning of wisdom.

The common-sense view of reasonable people is that they should be ever alert for anomalies that conflict with their current beliefs.

This is not very different from science, where common-sense observations and rational interpretations are accepted as fact unless and until they have been disproved.

In the everyday world, knowledge is a formalism of experience and is best acquired through trial and error and the subjection of 'facts' to iterative validation and invalidation. Science does not differ greatly from this, though experiment may play a greater part.

The main function of formal logic is to point out various fallacies. Most paradoxes are easily solved merely by recognising that the paradox statement is false.

Experiment is the cornerstone of science, but experiment is not always feasible, and theories may be built up as expected consequences and correlates of tested hypotheses. Such theories must remain more tentative than those that have been experimentally verified. Epidemiological studies relying entirely on statistics come into this category. Statistics is horribly misused, even by scientists.

Another approach where experimentation is not possible is to use models and simulations. Models are set up based either on retrospective data or theory or both and are testing in simulations using retrospective data. In meteorology, the success of this method can be judged by the accuracy of weather forecasts.

Most of our thought processes are associative and analogical rather than sequential logical.

Reason may not always be the best way to make a decision, if only because by the time you reach a decision the lion will have eaten you.

Physics remains the core science, though many recent concepts of the universe are not scientific in that they are not disprovable (this is not true of Einstein's theories, where several experimental tests have produced accordant results).

The cornerstone of biology is evolution, though the theory itself is continually evolving. However, it remains misunderstood by many.

The mind is a function of the brain and has no independent existence. Consciousness is also a brain function. The scientific study of the brain as such is largely the province of anatomy and neurology, while psychology probes the mind, and psychiatry fits somewhere in between. However, the study of mental effects is not always easy. Consequently, many of the conclusions of psychology are tentative at best, and pseudoscience abounds.

Medicine is progressively becoming more and more scientific, with controlled trials demanded of all new treatments and the requirement for all practices to be evidence based. As our knowledge of physiology expands, diagnosis more and more depends on laboratory tests and imaging. This is good to a point, but can result in many unnecessary tests being performed and in the tendency to label any anomaly as indicating, or even constituting, a disease. Frequently, also, psychological and sociological factors can be at least as important as biological ones. However, the acceptance of scientific medicine still has a long way to go and so-called alternative medicine, much of which is either fraudulent or misguided, continues to have a large following.

Economics is a 'soft science' which relies largely on models which, like those in meteorology, are largely retrospective and based on unproven theories.

Theories of business management are also largely untested.

Science is opposed by Antiscience, junk science, pathological science and pseudoscience. Perhaps the most pernicious of these is the Antiscience of academia fuelled by postmodernist concepts that everything is relative and there is no absolute truth.

New Age practitioners also have an entirely subjective view of truth (truth is whatever you want it to be). Evidence for their beliefs is entirely based on testimonials, placebo effect, wishful thinking, selective thinking, subjective validation, sympathetic magic, and communal reinforcement. The same can be said for those who believe in the paranormal, the occult, vital forces, the supernatural, ETs, UFOs and other strange phenomena, and mystics in general.

Religion is a subcategory of supernaturalism. Belief in God is irrational to the point of absurdity.

The argument that ethics and morals depend on religion is manifestly absurd.

Logical positivism sought to legitimise philosophical discourse by putting it on a basis shared with the empirical sciences. Its central thesis is that only statements verifiable through empirical observation are cognitively meaningful. This is rather in accord to my view of the sensible universe, but it needs to be remembered, as Karl Popper pointed out, that, to be truly scientific, an assertion must be capable of disproof. I firmly believe that the only way we can apprehend reality is through our senses and the only way we can establish the validity of our observations and their ascendants and consequences is through logic and rationality, the foremost form of which is the scientific method.

The humanities use methods that are primarily critical or speculative and have a significant historical element—as distinguished from the mainly empirical approaches of the natural sciences.

The historical method consists of the examination of primary sources and other evidence in an attempt to establish an internally and externally consistent account.

Pseudohistory purports to use the same methods but is intellectually inconsistent with the historical record and common-sense understanding and is based upon or derived from a theory which the work is intended to establish.

Geography is a science with important interactions with humanistic studies.

Languages may be studied scientifically or as part of the humanities.

Politics is difficult to study scientifically because of the difficulty of experiment.

Traditionally, education was concerned mainly with ensuring the survival (or, when this was already more or less assured, the happiness or achievement) of the individual, and with preserving the society. There has always been considerable tension between these two objectives.

As regards methods of teaching, the argument has largely centred on whether content or process should be emphasised.

The most pernicious facet of modern education is the influence of postmodernist teachers who believe that it is their job to produce ideal citizens for the ideal world they will produce.

One manifestation of this is 'progressive education' and its progeny 'productive pedagogy', which are full of untested (and largely untestable) claims and may masquerade as 'outcome-based learning' (a case of 'wolf in sheep's clothing').

The Sensible Universe

There is no intrinsic meaning of life beyond what we ourselves imbue it with.

www.ingramcontent.com/pod-product-compliance
Lightning Source LLC
Chambersburg PA
CBHW020548220526
45463CB00006B/2234